The Principles of Radiography

By

James Arnold Crowther

Published by Forgotten Books 2012

Originally Published 1922

PIBN 1000093007

THE PRINCIPLES OF RADIOGRAPHY

THE
PRINCIPLES OF
RADIOGRAPHY

BY

J. A. CROWTHER

M.A., Sc.D., F.Inst.P.

UNIVERSITY LECTURER IN PHYSICS APPLIED TO MEDICAL
RADIOLOGY ; DEMONSTRATOR IN PHYSICS IN THE
CAVENDISH LABORATORY, CAMBRIDGE

WITH 55 ILLUSTRATIONS

LONDON
J. & A. CHURCHILL
7 GREAT MARLBOROUGH STREET
1922

PREFACE

In this little volume I have tried to give an intelligible, though non-mathematical, account of the physical principles involved in the production of a radiogram, and in the construction and use of the apparatus employed for the purpose. The book contains little or nothing that an experienced X-ray worker will not have had to discover for himself during the course of his work. The discovery of physical principles with the aid of an elaborate X-ray installation is, however, apt to be an expensive as well as a tedious process, and the ordinary text-books of Physics are not primarily concerned with the needs of radiographers. It seemed, therefore, that practitioners and others who might be commencing this fascinating branch of work with no greater knowledge of physics than some hazy recollections of a first M.B. examination, might welcome a brief explanation of the principles of the subject on the physical side. With the medical side of the subject it is not within my province to deal.

The subject-matter of this volume formed part of a series of lectures given in connection with the recently established Diploma in Medical Radiology and Electrology, at Cambridge. The interest shown by the members of the class in the physical principles of the subject, and the difficulty experienced in finding any book which met our immediate requirements, encourages me to hope that these pages may prove of use to a wider circle.

I wish to express my thanks to Dr. Shillington Scales for the use of the radiogram from which Fig. 47 has been reproduced ; also to Messrs. Cuthbert Andrews for the illustrations of the X-ray tubes ; to Messrs. F. R. Butt & Co. for the illustrations of their couch and viewing box ; and to Messrs. Baird and Tatlock and the Cambridge and Paul Scientific Instrument Co. for the blocks of Figs. 3 and 11 respectively.

J. A. C.

CAMBRIDGE

CONTENTS

THE
PRINCIPLES OF RADIOGRAPHY

CHAPTER I

ELECTRICAL PHENOMENA

1. INTRODUCTION.—The discovery of X-rays by Rontgen in 1894 may be considered as one of the more recent advances of electrical science. Practically a century of painstaking and brilliant research, both practical and theoretical, into the properties of electricity had preceded it, and rendered it possible. Some acquaintance with electrical science is, therefore, almost indispensable for the proper understanding of the principles of radiology. It is true that so much thought and invention has now been applied to the construction of X-ray apparatus that almost any person of intelligence can work at any rate one of the smaller installations with reasonable success by merely following the instructions of the manufacturer. But it is equally true that scientific apparatus only yields its best results in the hands of one who understands the principles on which it acts. It is necessary, therefore, to preface our discussion of the production and properties of X-rays with a recapitulation of some of the more important facts of electricity.

2. ELECTRICAL ATTRACTIONS.—If a stick of ebonite or sealing wax is rubbed with flannel it is found to attract light objects such as feathers, scraps of paper, or pith balls. A glass rod rubbed with silk produces similar effects. These attractions are called electrical attractions, and the rods are said to be electrified by friction, or to have acquired a charge of electricity. The name " electricity " was given to the phenomenon by

I

William Gilbert (1603), but the effect itself was known to the Greek philosophers. In spite of the immense amount of electrical research, we know very little more as to its real nature than they did.

Although both a rubbed glass rod and a rubbed ebonite rod attract light objects, there is a difference between the electricity excited in the two cases. If an ebonite rod is rubbed with flannel and suspended on a stirrup so as to be free to turn, and a second ebonite rod, electrified in the same way, is brought near it, it will be found that the two rods repel each other. If, however, the charged glass rod is brought near the suspended ebonite rod, the latter will be attracted. Similarly an electrified glass rod is repelled by a second charged glass rod, while it is attracted by a charged ebonite rod. There are thus two kinds of electricity, each of which repels other electrical charges of the same kind as itself, while it attracts charges of the other kind. This is usually expressed by saying that like charges repel ; unlike charges attract.

The kind of electricity excited on a glass rod by rubbing it with silk was called vitreous, but is now known as positive electricity ; while the kind excited on an ebonite rod by rubbing it with flannel was called resinous, but is now always called negative electricity. The terms " positive " and " negative " arose from the fact that if equal quantities of vitreous and resinous electricity are placed on the same body, the latter shows no signs of electrification. The result of adding together equal quantities of vitreous and resinous electricity is thus zero, and this result can be expressed mathematically by assigning a positive sign ($+$) to the one and a negative sign ($-$) to the other. It is therefore a convenience in the mathematical theory of electricity to regard the two kinds of electricity as being of opposite sign.

3. CONDUCTORS AND INSULATORS.—If a brass rod is held in the hand and rubbed with silk it shows no signs of electrification. If, however, we mount the brass rod on an ebonite handle, and holding it by the handle so that no part of the rod is in contact with the hand, flick it gently with the silk, the brass becomes positively electrified. If now we touch the brass

with the hand the charge disappears. The electricity passes through the hand and the body to the ground. Similarly if the brass rod is touched with a metallic wire, or with a damp cloth, it loses its charge. If, however, the rod is touched with ebonite, rubber, sulphur, wax, or dry silk, the charge is retained. We can thus divide substances into two classes, according as they do, or do not, permit the passage of electricity through them. The former are called conductors; the latter non-conductors, or, more usually, insulators.

The distinction between insulators and conductors, though often very marked in practice, is really only one of degree. The best insulators allow electricity to leak through them slowly, while the best conductors afford a certain amount of resistance to its passage. All metals conduct, silver and copper being the best conductors. Solutions of salts in water also conduct well. The human body is, therefore, a conductor. Sulphur, ebonite, shellac, gutta-percha and rubber are among the best insulators. Hard woods are partial insulators, but it is inadvisable to rely on their insulating properties. Air is, of course, one of the best of insulators under ordinary circumstances.

4. THE NATURE OF ELECTRICITY.—Although we cannot answer the question " What is electricity ? " we know something about its structure. It has been shown that negative electricity consists of very minute particles, all of exactly the same size, charge and mass. Negative electricity thus is atomic in structure, and the charge on one of these particles is the smallest charge which we can have. These negatively charged particles, which are known as *electrons*, are exceedingly minute. Their mass is about 1/1800th of the mass of a hydrogen atom, and their radius is only 1/10,000th of that of an atom. The size of the electron, therefore, bears to the size of an atom about the same relation as the size of a pea to that of a cathedral.

These electrons are contained in the atoms of all substances. In the case of insulators they are apparently unable to leave the atom unless acted upon by very considerable forces. The atoms of conductors, however, appear to be able to give off a certain number of electrons spontaneously, by a process which we may regard as very similar to that of evaporation, so that a conductor always contains a considerable number of free electrons. As the spaces between the atoms even of a solid are large compared to the size of an electron, the free electrons are able to move through the solid in much the same way that the molecules of a gas make their way through a pile of lead shot.

Since every kind of matter contains electrons, an uncharged body must be one which contains equal and opposite charges of positive and negative electricity. We are not quite so familiar with the nature of positive electricity as with the properties of the electron. Positive electricity never occurs associated with a mass less than that of a hydrogen atom, and we have strong reasons for believing that, if we could extract all the electrons from any atom, the remainder would be simply positive electricity. It is believed that a hydrogen atom consists of one negative electron and one positive charge. For further information on these points the reader may refer to the author's "Molecular Physics." In any case, the positive electricity seems unable to leave the atom and can, therefore, only move when the latter moves. Conduction through solids is, therefore, due entirely to the motion of the negative electrons. This result has been verified experimentally.

A body is said to be insulated if it is completely surrounded by non-conductors and so is in a condition to retain any charge which is placed upon it. Since air is a nearly perfect insulator, a body will be insulated if it is supported entirely by non-conductors. A body connected by conductors to the ground is said to be earthed. The body, the conducting connection and the earth then form one continuous conductor. The electricity spreads itself over the whole, and as the earth is very large compared with the electrified body, the fraction of the charge which remains on the body after earthing is practically *nil*. In order that the body may be efficiently earthed it is best to connect it by a copper wire to the nearest water pipes. Failing these, it can be connected to a large plate of metal buried in moist soil.

5. Law of Force between Electrical Charges.—The force with which one electrical charge attracts or repels another falls off rapidly as the distance between the charges is increased. Exact experiments show that the force is inversely proportional to the square of the distance between the two charges. It also depends on the medium between the charges. Thus, the force between two given charges is only half as much in paraffin oil as it is in air. These results are expressed in the formula—

$$F = \frac{1}{K} \cdot \frac{qq'}{d^2}$$

where F is the force exerted by either of the charges on the other, q and q' are the strengths of the two charges, d is the distance between them, and K is a constant which depends on the medium separating the charges, and is known as the specific inductive capacity of the medium. Its value is taken as unity for air. The specific inductive capacity of glass is about 6, that of sulphur 4, and that of paraffin oil about 2. The greater the specific inductive capacity, the smaller the force.

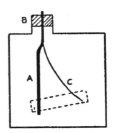

Fig. 1.—A simple electroscope.

6. Measurement of Electrical Charges. —The fact that electrical charges attract or repel each other according to definite laws can be used to detect, and with proper arrangements to measure, charges of electricity. The simplest device for this purpose is the gold leaf electroscope. The necessities of modern research have brought this old-fashioned instrument into prominence again, and various improved varieties have been devised. A simple, but for many purposes quite efficient, type is shown in Fig. 1. It consists of an outer case, generally rectangular in shape, made of metal and earthed. A fairly stout metal rod A passes through the top of the case and is insulated from it by a plug B of some efficient insulating substance, preferably sulphur, or ebonite. The portion of the rod inside the case is beaten out into a flat plate and the gold leaf C is attached

to its upper end.   Glass windows are made in the front
and back of the case, so that the movements of the gold
leaf can be observed.   When the rod is uncharged the leaf
hangs vertically downwards.   If, however, a charge is given to
the rod, say, for example, by touching it with a charged ebonite
rod, the rod and the gold leaf become charged with electricity
of the same sign, and the latter is repelled, as shown in the
figure.   The greater the charge, the greater will be the deflexion
of the leaf.   The motion of the leaf can be measured by means
of a paper scale pasted on the back window, or more accurately
by viewing the gold leaf through a long focus microscope having
a divided scale in the eyepiece.   With the latter arrangement
the instrument becomes capable of considerable accuracy, and
is largely used in radioactive measurements.

7. ELECTROSTATIC INDUCTION.—If a charged conductor is
brought into contact with an uncharged conductor, the former
imparts some of its charge to the latter, which thus becomes
charged with electricity of the same kind as the former.   Experiments show that the total charge remains the same, but it is
divided between the two con-
ductors in proportions which
depend on the relative sizes and
shapes of the two.   This process
is called charging by conduction.

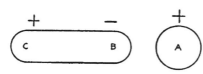

FIG. 2.—Electrostatic induction.

We can, however, produce a
charge on the uncharged conductor without the charged
conductor losing any of its charge.   Suppose a positively
charged metal sphere A (Fig. 2) is brought near one end of
an uncharged insulated conductor BC.   It will be found that
the end B nearer to A acquires a negative charge, while the
opposite end C acquires a positive charge, the two charges
being equal and opposite.   If now the conductor BC is earthed
for a moment, it is found that the positive charge flows to earth.
The negative charge, however, remains on the conductor.
Thus when the earth connection is removed, the conductor BC
is left with a negative charge, which remains on the conductor
(which is, of course, now insulated), even after the charge A is
withdrawn.   We have thus produced a negative charge on the

insulated conductor without altering in any way the amount of the charge on A. This process is called "charging by induction."

A convenient way of considering the matter is to suppose that the uncharged conductor contains in its uncharged state equal charges of positive and negative electricity. When the positive charge A is brought near the conductor it attracts electricity of the opposite sign, and repels the electricity of the same sign to the opposite end. On earthing the conductor the latter is repelled to earth, leaving the conductor with a negative charge.

The amount of charge induced on BC depends on the magnitude of the charge on A and the nearness of A to BC. The nearer A is brought to BC the larger will be the resultant effect. If, instead of using a conductor such as BC, we employ a hollow metal can, and lower the charged sphere A into it, being careful not to let it touch the sides of the can, the charges induced on the latter will be exactly equal to the inducing charge. This result, which is due to Faraday, has considerable bearings on electrical theory.

Very nearly the same effect is obtained if the two conductors consist of large, flat plates, separated by a very small thickness of some insulating substance. This is the principle of the instrument known as the electrophorus. The electrophorus consists of a flat cake of ebonite, or of a mixture of various resinous substances, which is electrified by rubbing it with flannel.

A slightly smaller flat brass disc, carried by an insulating handle and having its lower face coated with a non-conducting varnish, is placed on the cake. Since the latter is negatively charged, a positive charge is induced on the lower face of the disc, and a negative charge on the upper face. If the disc is momentarily earthed, the latter flows to earth, leaving the disc with a positive charge. The disc can now be removed and its charge transferred to some other conductor. As the negative charge on the cake is not affected in any way, the process can obviously be repeated until the cake gradually loses its charge owing to defective insulation.

8. INFLUENCE MACHINES.—By suitable mechanical devices
the necessary succession of operations for producing a charge
by induction can be carried on continuously by simply turning
a handle.   The arrangement then forms what is known as an
influence machine.

The best known of these is the Wimshurst.   The Wimshurst
machine (Fig. 3) consists of two insulating plates, furnished with
a series of conducting metallic sectors on their outer surfaces.
The plates are mounted close to each other and geared so that
they rotate in opposite directions when the handle of the machine

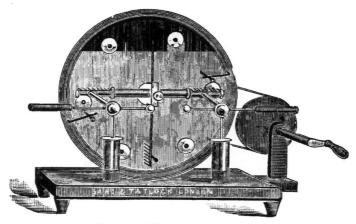

FIG. 3.—Wimshurst machine.

is turned.   Each plate is furnished with a long conducting rod
which makes contact, by means of metallic brushes, between
opposite sectors on the plate.   The two rods are mounted at
right angles to each other and approximately at an angle of
45 degrees with the horizontal.   At opposite ends of the horizontal diameter of the plates are the collectors, or combs, which
consist of a number of sharp points projecting close to the
rotating sectors on each side of the plates.   The effect of these
points is to collect the charges which are on the sectors and to
transfer them to the insulated brass knobs which form the
poles of the machine.   To start the machine a charged rod is
held opposite one of the brushes, and the handle is turned.
The poles begin to charge up with electricity of opposite sign.
The charged rod can then be removed, and the machine will

continue to produce charges as long as the handle is turned. If the knobs are not too far apart a succession of sparks will pass between them, owing to the tension of the electricity produced being sufficiently high to break down the insulating power of the air. If the knobs are too far apart for this, the electrical charges will discharge into the air in the

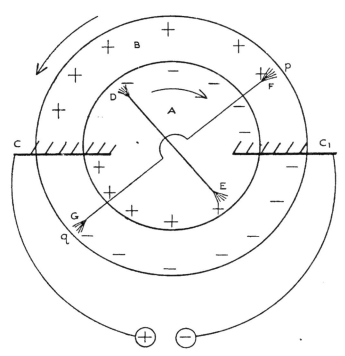

FIG. 4.—Principle of the influence machine.

form of a luminous brush or leak away across the insulating supports.

The action of the Wimshurst will be understood from the diagram in Fig. 4. A and B represent the two plates, rotating in the direction of the arrows; C and C' are the two collectors, and DE and FG the brushes with their connecting rods. Suppose now that a charged ebonite rod is placed opposite the brush F. The brush and the sector p in contact with it become positively charged by induction, while the brush G and the sector q become negatively charged. The charged sectors are carried round by the revolution of the plates until they are brought opposite to the ends D and E of the other connector. Here they act inductively on the connector, inducing a negative charge at D and a positive charge at E. These charges are carried away in turn by the sectors on A and, coming opposite the ends of the connector FG, induce new charges on F and G of the same sign as before. The charged rod can now be removed, as the induction on FG is maintained by the charges on the sectors coming from D and E. In the same way the charges on D and E are maintained by induction from the charged sectors coming from F and G. The machine, when once started, will continue to work as long as the handle is turned.

After acting inductively on the brushes the charged sectors are brought opposite to their respective collectors, where their charges are transferred, by the action of the points, to the terminals of the machine. It will be seen that all the sectors passing the collector C are positively charged, and all those passing C' are negatively charged. C and C' are thus respectively the positive and negative terminals of the machine.

The influence machine has the advantage of being quite self-contained, and can, therefore, be used where other sources of supply, such as the mains of an electric power supply company, are not available. It produces electricity at a high tension, very suitable for working X-ray tubes, and for some kinds of electrical treatment. In practice, however, it has not been found in the past to give satisfactory results in the British climate. This seems to be due principally to the difficulty of securing adequate insulation of the various parts. The water vapour present in large amounts in the atmosphere condenses on the surface of the plates and on the insulating supports, forming a slightly conducting film over them, which allows the electricity to leak away. As the total quantity of electricity produced by the machine is not large, any loss of this kind is very serious.

Various attempts have been made to improve the machine. At one time a completely enclosed machine, working in compressed gas, had some vogue, but did not prove altogether satisfactory.

A new type of machine, with three or more pairs of rotating plates of ebonite, and driven at high speed by a motor, has recently been placed upon the market and appears to work satisfactorily except in very bad weather. A really efficient influence machine would certainly be a great convenience in X-ray work, as the current supplied is much more suitable for exciting X-ray tubes than that furnished by either high tension transformers or induction coils, and its independence of an external electrical supply would be of great service in remote districts.

9. POTENTIAL.—We have seen that if a positively charged body is placed in contact with an uncharged body, the former will give a portion of its charge to the latter. In general, if any two positively charged bodies are placed in contact, electricity will pass from one to the other. It is by no means always the conductor having the larger charge which gives some of its charge to

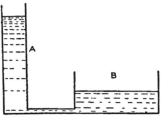

FIG. 5.—Illustration of electrical potential.

the other. It is quite easy to arrange the conditions so that the conductor with the smaller charge parts with electricity to the conductor which has already the larger charge. On the other hand, when the conductors have shared their charges, no further transference of electricity takes place between them. Each has a definite charge, which is not altered by leaving them in contact. These two charges are not usually equal.

What property determines the direction in which electricity will flow when two charged bodies are joined by a conductor ? We can, perhaps, make the matter clearer by considering the analogy between electricity and a fluid. If we compare a quantity of electricity to a quantity of water, a conductor may be compared to a pipe which allows the water to flow through it. Now if we have two vessels, such as A and B (Fig. 5), connected by a pipe, we know that the water will flow along the pipe from the vessel in which the level is higher to the one in which it is lower quite irrespective of the quantities of water in each vessel. Thus, if A is a narrow vessel and B a wide one, the quantity of water in B may easily exceed that in A. Nevertheless, water will flow from A to B until the level is the same in the two vessels.

The corresponding property in electricity, which determines the direction in which electricity will flow, is called electrical potential, or electromotive force. Thus we may define potential as the electrical condition which determines the direction in which electricity will flow.

A still closer analogy is that between potential and the pressure of a gas. We have considerable reason for supposing that the negative electrons in a conductor are in a condition very similar to that of the molecules of a gas. If two gas bags, containing gas at different pressures, are connected by a pipe, the gas will flow from the bag where the pressure is high to the one where it is low until the pressure is the same throughout. We may thus regard potential as measuring the electrical pressure of the charge.

The theory of potential was worked out before the nature of the electrical charges was understood. As a matter of convenience, it was decided to consider all the changes as being brought about by a flow of positive electricity. This was unfortunate, as we now know that, at any rate when the transference takes place through metals, it is the negative electricity which really moves. It is, however, too late to alter the nomenclature of the subject, and no trouble will arise if it is remembered that a negative charge will always move in the opposite direction to that of a positive charge. Thus when a positively charged conductor is earthed, the conductor loses its charge. This is explained by supposing that the positive electricity flows to earth, and the conductor is said to be at a higher potential than that of the earth. On the other hand, when a negatively-charged conductor is earthed it is assumed that sufficient positive electricity flows from the earth to the conductor to neutralise the charge upon it, and the negative conductor is said to be at a lower potential than that of the earth. The potential of the earth is generally taken as zero, so that any earthed conductor is at zero potential. The potential of the negatively-charged conductor in the example would, therefore, be said to be negative, as it is less than zero.

Electricity will not flow from one place to another at the same potential, no matter what the charges may be. Thus, in the induction experiment in § 7, one end of the conductor has a positive charge, the other end a negative charge, but the two charges do not recombine, although they are situated on the surface of a conductor, because the potential of the conductor is the same throughout. When the conductor is earthed at any point positive electricity flows to earth, because the potential of the conductor is positive. It will be noticed that the negative charge on the conductor we are considering is at a positive potential. Thus potential is simply that condition which determines the direction in which positive electricity will flow.

The difference of potential between two points is measured numerically by the work which would have to be done to transfer a unit quantity of positive electricity from the point at the lower to the point at the higher potential. By the principle of the conservation of energy this work will be returned if the unit quantity is allowed to flow from the higher to the lower potential. Thus, when electricity flows from a high to a low potential, energy is liberated which can be utilised in various ways; for example, to turn a motor, heat a wire, or to decompose a chemical compound. The amount of work so liberated is equal to the quantity of electricity multiplied by the difference of potential through which it falls.

The meaning of this definition of difference of potential will be clearer if we consider again the analogy with water level. The difference in level between two water surfaces is, in practice, measured in feet or metres. Since, however, the work done in raising unit mass of water from one level to the other is proportional (at any one place on the earth's surface) to the difference in level, we might have defined the difference in level by the work which would have to be done to transfer unit mass of water from one level to the other. This would correspond exactly to the electrical definition of difference of potential. In both cases the work done is stored as potential energy.

10. Practical Measurement of Potential.—Potential can be measured in practice by the gold-leaf electroscope. In fact, the movement of the leaf is due to the fact that there is a difference of potential between it and the case of the instrument, and the deflection of the leaf is a measure of the potential of the gold-leaf system. The use of the electroscope for measuring charges depends on the fact that for a given conductor, such as the gold-leaf system, the potential to which it is raised is directly proportional to the charge which is put upon it. The deflection is thus proportional to the charge put upon the system.

To measure the potential of a conductor, we connect it by a

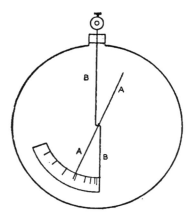

conducting wire to the projecting end of the gold-leaf support, and note the deflection of the leaf. The conductor, the wire, and the gold leaf are now all part of one conducting system. They must, therefore, all be at the same potential as soon as the electricity has come to rest, for otherwise the electricity, being on a conductor, would continue to flow as long as there was any potential difference between the different parts of the system.

Fig. 6.—Braun electrometer.

The potential of the conductor is thus the same as the potential of the gold leaf, and is measured by the deflection of the leaf. The gold-leaf electroscope does not give us directly the value of the potential in terms of the units we have just denned. If we wish to know the absolute value of the potential it is necessary to calibrate the instrument by applying different known potentials to it and noting the corresponding deflections. For many experiments, however, comparative measurements are quite sufficient.

As the gold leaf is very liable to breakage, other instruments of a less fragile nature have been designed. The Braun electrometer (Fig. 6) consists of a narrow vertical brass plate B, suspended inside a circular metal case, which is earthed, and insulated from it by a plug of ebonite or amber. A light aluminium needle AA is pivoted at the centre of the plate, on an axis just above its centre of gravity. When the plate and needle are charged the needle is repelled in exactly the same way as the gold leaf in the gold-leaf electroscope, the deflection of the needle being measured on a circular scale. The scale is generally graduated in volts. (A volt is the practical unit of potential difference, and is equal to 1/300th of the absolute unit of potential defined in the previous section.)

The Kelvin electrostatic voltmeter consists of a vertical aluminium plate A pivoted on a horizontal axis just above its centre of gravity. Two metal plates, Q, shaped as shown in the diagram (Fig. 7), are placed one on each side of the needle, and are in metallic contact with each other but insulated from the needle. The needle itself is connected to the case. If the needle and the plates are at a difference of potential the needle is attracted and moves into the space between the plates, its motion being read off by means of the pointer p, which travels over a circular scale. The scale can be graduated empirically, to give the potential difference in volts.

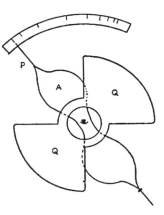

FIG. 7.—Kelvin electrostatic voltmeter.

A more sensitive form of the same instrument has a considerable number of aluminium plates, similar to A mounted on the same axle. Each of these moves between a pair of quadrantal plates similar to Q, all the plates being in metallic contact. The principle is the same as that of the electrostatic voltmeter already described, but, owing to the large number of plates, the deflection for a given potential difference is greatly increased. It is usual to construct this type of instrument, called the Multicellular voltmeter, with the plates horizontal and the axis vertical. The moving plates are controlled by a fine spring instead of by gravity as in the simpler type.

11. ELECTRICAL CAPACITY.—If a positive charge is given to

an uncharged conductor the potential of the latter is increased. It is found that for a given conductor the potential is directly proportional to the charge placed upon it. The ratio of the quantity of electricity on the conductor to the potential of the conductor is called the capacity of the conductor. Thus, if Q is the charge on the conductor and V its potential,

$$Q = CV,$$

where C is a constant for the given conductor and is called its capacity. The capacity of a conductor depends upon its size and shape. It also depends on the position of other conductors in its neighbourhood. The capacity of a conductor is greatly increased by placing an earthed conductor in close proximity to it. An arrangement by which the electrical capacity of a conductor is increased in this way is called an electrical condenser.

The action of a condenser can be illustrated very simply, using a pair of parallel plates. One of the plates is earthed; the other is insulated and connected to an electroscope. The plates are placed parallel to each other and some little distance apart, and the insulated plate is given a charge. If the earthed plate is moved gradually nearer to the charged plate it is found that the deflection of the gold leaf is gradually diminished, showing that the potential of the charged plate is being reduced. If the two plates are again separated the gold leaf returns to its original reading, thus showing that the charge on the plate has not been affected. As the charge remains constant while the potential is decreased, it is evident that the capacity of the insulated plate has been increased by moving the earthed plate nearer to it. It can be shown that the capacity of a parallel plate condenser is (*a*) directly proportional to the area of the plates; (*b*) inversely proportional to the distance between the plates; (*c*) directly proportional to the specific inductive capacity of the medium between them. The capacity of a parallel plate condenser is equal to $\frac{K\,A}{4\pi d}$, where A is the area of one of the plates, K the specific inductive capacity of the insulating substance between them, and *d* is their distance apart.

Condensers of very large capacity can be made by making a pile of alternate sheets of tinfoil and waxed paper, or better still mica. The sheets of tinfoil project alternately to left and right as shown in Fig. 8, the projecting portions being connected together. The arrangement forms a pile of parallel plate condensers, the waxed

FIG. 8.—Multiple plate condenser.

paper (indicated by the broken lines in the figure) serving as the insulation. The capacity of the system is equal to the sum of the capacities of the separate condensers into which it can be divided. In the figure there are six of these. As the distance between the plates is very small, the capacity of condensers of this kind may be very large. They cannot be used for high potentials, however, as a potential difference of more than a few hundred volts would be sufficient to cause a spark to pass through the waxed paper, and thus ruin the insulation.

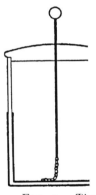

FIG. 9.—The Leyden jar.

For high potential work a condenser with a much greater thickness of insulation is required in order to withstand the strain. The greater distance between the plates, of course, very much reduces the capacity of the condenser. The usual form of condenser for high potentials is the Leyden jar. This consists of a glass jar (Fig. 9) closed by a hard wood stopper. The inside and outside are coated with tinfoil, which extends rather more than halfway up the jar. The upper portion of the jar is coated with shellac varnish to improve the insulation. Contact is made with the inner coating by means of a brass rod passing through the stopper. The outer coating is earthed. If a large capacity is required a number of these jars can be connected in parallel (as it is called) by joining together their inner terminals with copper wire. The capacity of the system is then the sum of the capacities of the separate jars.

The action of an electrical condenser can easily be understood by taking the case of two parallel plates.   Suppose the insulated plate is positively charged and the earthed plate is placed near it ; a negative charge will be induced on the latter which, if the two plates are very close, will be practically equal to that on the positive plate.   The forces which the two plates exert on a unit positive charge placed at a little distance from the plates are, therefore, almost equal in magnitude and opposite in direction.   Thus the resultant force would be almost zero, and very little work would have to be done in bringing up the unit charge to the insulated plate compared with the work which would have had to be done in the absence of the earthed plate.   In other words, the potential of the insulated plate is lowered by the presence of the earthed plate, and thus its capacity is increased.

The quantity of electricity which can be placed on a condenser is limited by the strength of the insulating medium, or dielectric, as it is called, between the plates.   If the potential difference rises too high, the medium breaks down and a spark passes, which discharges and, at the same time, ruins the condenser.   The potential difference required to send a spark through a centimetre thickness of the medium is called its dielectric strength.   The dielectric strength of air at atmospheric pressure is about 30,000 volts per centimetre, that of waxed paper is about 500,000 volts per centimetre, and that of glass from 300,000 to 1,500,000 volts per centimetre.   Mica has about the same dielectric strength as glass.   These figures are for parallel plates.   A very much smaller potential difference will suffice to produce a spark between pointed electrodes.

CHAPTER II

12. ELECTRIC CURRENTS.—If two conductors at different potentials are connected together by a conducting wire, electricity will flow along the wire from the one to the other until they are both at the same potential. If we are dealing with two insulated conductors, this equalisation will take place practically instantaneously. If, however, we connect the two conductors to the two poles of a Wimshurst machine and keep on turning the handle, the charges produced by the machine will pass into the conductors and thus tend to maintain the difference of potential between them. Under these circumstances it is evident that there will be a continuous flow of electricity along the wire. We may call this flow an electric current.

Although the difference of potential produced by an influence machine is often very high, the quantity of electricity generated is very small. Thus the current along a wire joining the poles of a Wimshurst is only small. Much larger currents can be produced in other ways. If a sheet of copper and a sheet of zinc are dipped in dilute sulphuric acid, it is found that there is a difference of potential between the two plates, the copper plate being at a higher (positive) potential than the zinc. Compared with the potentials we have been dealing with in the previous chapter the difference is very small—about 1·4 volts. If, however, we join the two plates by a conducting wire, the quantity of electricity flowing through the wire in a given time is much greater than that produced even by a large multiple plate Wimshurst. Thus, while the copper and zinc plates only produce electricity at a low potential, they produce it in far greater quantities than does the Wimshurst.

The arrangement of the two metallic plates in acid is called a voltaic cell.   The copper plate forms the positive pole, the zinc plate the negative.   The current is, by convention, said to flow from the positive to the negative pole, that is from the copper to the zinc, along the wire joining the two.   That is to say, it has been agreed to look upon the current as being a flow of positive electricity.   The convention is somewhat unfortunate, for, as we have seen, the current through the wire must be due to the flow of negative electrons along it, and these will, of course, move from the negative to the positive pole.   The convention is too firmly rooted to be displaced.

The very simple type of cell described does not work satis-

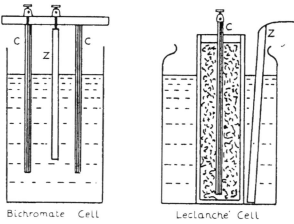

Bichromate   Cell   Leclanche´ Cell

FIG. 10.—Voltaic cells.

factorily owing to the deposition of hydrogen on the copper plate.   Several forms of voltaic cell have been devised in which this difficulty is overcome.   Of these the bichromate cell is the best for producing large currents.   It consists of a pair of carbon plates C C (Fig. 10), connected together outside the cell and forming the positive pole of the battery.   The negative plate Z is of zinc and is fastened between the two carbon plates. The liquid in the cell consists of strong sulphuric acid and potassium bichromate (which gives its name to the battery). The bichromate oxidises the hydrogen as fast as it is liberated, and thus keeps the cell in working order.   The liquid is highly corrosive, and some arrangement is generally added for removing the zinc plate from the liquid when the battery is not actually in use. Very large currents (up to 30 amperes) can be obtained from a bichromate battery, and it is useful where other sources of supply are not available.

Where large currents are not required, and especially for intermittent work, a Leclanche cell (Fig. 10) is often used, as it contains no corrosive fluids, and requires no attention except for the addition from time to time of a little water. The posi-:ive plate is a plate of carbon C, which is packed round with a ιixture of manganese dioxide and coke, usually contained in a porous pot. The negative plate Z is of zinc dipping into a saturated solution of sal ammoniac in water. The " dry " cells on the market are practically Leclanche cells in which the sal ammoniac is mixed with sufficient damp plaster of Paris and flour to render the cell unspillable.

For practical purposes, however, current is generally produced by means of elaborate machines, known as dynamos, as electricity can be generated on a large scale much more cheaply by a dynamo than by the use of a voltaic battery. The dynamo can be connected by wires to terminals in the buildings where the current is required. If these terminals are tested when the dynamo is working, it will be found that there is a difference of potential between them ; the terminal at the higher potential is termed the positive terminal, the other is called the negative terminal. When the terminals are connected by a wire an electric current passes along the wire from the positive to the negative terminal.

13. EFFECTS OF THE CURRENT.—The effects produced by an electric current fall into three main classes : (a) magnetic ; (b) thermal ; (c) chemical.

Magnetic Effects.—If a wire carrying a current is held just above a magnetic compass needle, the latter will be deflected. The deflection is a maximum if the wire is held parallel to the direction in which the needle was originally setting, that is parallel to the magnetic meridian. The amount of the deflection depends, among other factors, on the strength of the current ; the greater the current the greater the deflection. The direction in which the north-seeking pole of the compass needle will be deflected can be found by a rule given by Maxwell. Let us suppose that we are screwing a corkscrew into a cork in the direction in which the current is travelling. Then the north-seeking pole of the needle will be urged in the direction taken by the thumb in turning the corkscrew. Thus if the wire is held in the meridian above the needle, and the current is travelling from south to north, the north-seeking pole will be deflected towards the west ; if the current is travelling from north to south, it will be deflected towards the east. This rule enables us to find the direction in which the current is flowing in a given circuit.

If the wire is wound in a spiral on a circular cylinder a magnetic field is set up inside the spiral, and a piece of iron

Fig. 11.—Interior of moving coil ammeter.

placed inside the spiral becomes strongly magnetised while the current is flowing. If the iron is what is known as " soft " iron the magnetism practically vanishes when the current is stopped. The " primary " of an X-ray coil, or transformer, consists of a spiral of thick copper wire wound on a core of soft iron. In the case of the coil the iron core is straight. In the case of the transformer it forms a continuous ring of metal.

Conversely, if a coil of wire is suspended so that it can turn freely about its supports and placed between the poles of a horse-shoe magnet, it will be deflected when a current is passed through it, the turning force increasing with the strength of the current. The instruments used for measuring currents, and known as ammeters, generally work on this principle. The coil (Fig. 11) is suspended by flat springs, like the hairspring of a watch, through which the current enters and leaves the coil. A light pointer rigidly attached to the coil moves over a graduated scale, so that its deflections can be easily observed. The scale is generally graduated to read off the strength of the current directly.

Thermal Effects.—A current passing through a wire generates heat in the wire. If the wire is very thick and made of good conducting material, the heat generated will be comparatively ıall. If, however, the wire is thin and the current strong, the heat produced may be sufficient to raise the wire to incandescence. This is the principle of the electric filament lamps. This production of heat is an invariable accompaniment of the electric current, and takes place in our X-ray coils, measuring instruments, resistance regulators, and so on, whenever the current is passing. The apparatus is designed by the makers to carry a certain maximum current without damage. If this maximum current is exceeded the instrument may be ruined by the heat produced. To avoid accidents of this kind every circuit should contain a fuse. This consists of a piece of easily fusible wire of such a thickness that the heat produced in it will be just sufficient to melt it when the maximum current is reached. Thus if the maximum current which the apparatus will safely carry is accidentally exceeded, as sometimes happens in practice, the fuse will melt or " blow," as it is called. This at once breaks the circuit and stops the current, thus protecting the more valuable parts of the apparatus from damage. The fuse is constructed so that a fresh piece of fuse wire can easily be inserted when required. An examination of the fuses is always the first step called for in the case of any sudden stoppage of the current.

Chemical Effects.—If the wire carrying the current is cut, and the two ends dipped into a beaker of dilute sulphuric acid, bubbles of gas are seen streaming off from the ends of the wire. These are formed by the decomposition of the solution by the current, the gas evolved at the wire connected with the positive pole being oxygen, and that at the negative pole hydrogen.

If the current is passed through a solution of sodium sulphate in water, sodium is liberated at the negative pole, and the acid radicle at the positive pole. The sodium reacts with the water to form sodium hydrate, while the acid radicle reacts to form sulphuric acid. Thus, if the solution is coloured with litmus, the litmus round the positive pole will be turned red, and that round the negative pole blue. This affords a very convenient method of determining which of the terminals on a switch-board is positive, and which negative. A piece of blue litmus paper is soaked in a solution of sodium sulphate and laid across the terminals. The portion of the paper in contact with the positive pole will be turned red.

Similarly, if a piece of starch paper moistened with potassium iodide is placed across the terminals, iodine will be liberated at the positive terminal, producing a blue stain.

14. STRENGTH OF THE CURRENT.—We have agreed to regard an electric current as being due to the flow of electricity along the wire from the positive towards the negative end of the wire. It is thus in many ways analogous to the flow of water through a pipe. In the latter case we measure the strength of the current by finding the number of gallons of water which flow across any cross section of the pipe in a given time, say one second. In the same way we can regard the strength of the current as the quantity of electricity which passes across any cross section of our conductor in one second.

In practice the strength of an electric current is measured by the magnetic effect which it produces. The practical unit of current is called the *ampere*. Instruments for measuring current are known as *amperemeters*, or more briefly *ammeters*. We have already described the principle on which these instruments work (§ 13). The current to be measured is passed directly through the instrument, care being taken that the positive end of the wire is joined to the positive terminal of the instrument, which is marked, either with a little + sign, or occasionally by being painted red. If the current is passed the wrong way the instrument may be damaged. A modern ammeter is a fairly complicated piece of apparatus, and if damaged will in general require the services of a competent instrument maker to put it right.

When the current is small, as in the case of the current through an X-ray tube, a smaller unit equal to one-thousandth of an ampere is employed. This is called a milliampere, and is measured by instruments working on the same principle as the ammeter, but arranged to give a deflection with much smaller currents.

15. CURRENT, CHARGE, AND POTENTIAL.—A complete electric circuit may be regarded as in many ways analogous to an hydraulic system. In an hydraulic system water is pumped into a tank at a high level, and is then allowed to flow through pipes to some reservoir at lower level, the energy in the water being utilised on the way to drive various kinds of machinery, operate lifts, and so on. In the same way we may regard a voltaic battery or a dynamo as a kind of electric pump, raising electricity from a low to a high potential, the positive terminal corresponding to the high level reservoir, and the negative to the low level. When the two are joined by a wire, electricity flows from the positive to the negative terminal, and its potential energy is converted into work and can be utilised in various ways.

Just as the work which can be done by unit mass of water is proportional to the height through which it falls, so the work done by unit quantity of electricity is proportional to the potential difference between the ends of the conductor. Now the total quantity of electricity flowing from one terminal to the other in a given time is obviously equal to the strength of the current multiplied by the time for which it flows. Thus, if a current of C amperes flows for t seconds, the total quantity of electricity Q passing any point on the conductor is given by

$$Q = C.t.$$

The practical unit of quantity is the coulomb, and is the total charge passing any given point on the conductor when a current of 1 ampere flows for one second.

Now the work which can be done by a quantity of electricity O, falling through a difference of potential V, is proportional to the product O.V. Thus if a current C is flowing between two terminals, having a potential difference V, the work done by the current in a time *t* is proportional to Q . V, that is to C . *t* . V. The work done when a current of 1 ampere flows for one second between two points at a difference of potential of one volt is known as a joule.   One joule is equal to ten million ergs.   Thus we may write the work done by the current

$$= C . t . V \text{ joules}$$
$$= C . t . V \times 10^7 \text{ ergs.}$$

16. ELECTRICAL POWER.—The power of an electric circuit is measured by the rate at which it can do work, that is to say by the work done divided by the time taken.   Since the work done is equal to C . V . *t*, the power is obviously equal to C . V, that is, the product of the current into the difference of potential.

If the current is measured in amperes and the potential difference in volts, the power is measured in a unit known as a watt.   Thus

$$\text{amperes} \times \text{volts} = \text{watts.}$$

It is found that 746 watts are equal to 1 H.P.

Thus if we pass a current of 5 amperes through an induction coil at a pressure of 200 volts, the power supplied to the coil is $5 \times 200 = 1,000$ watts, or 1 kilowatt.   This is equal to about $1\frac{1}{3}$ H.P.   To supply the same power with a pressure of 50 volts would require 20 amperes.   Again, if an X-ray tube is taking a current of 3 milliamperes at 150,000 volts, the power supplied to the tube is $\frac{3}{1000} \times 150,000$ watts, that is, 450 watts, or over $\frac{1}{2}$ H.P.

17. PRODUCTION OF HEAT BY A CURRENT.—If no arrangements are made for utilising the energy of the current, if for example the current is simply allowed to flow along a conducting wire, the whole of the energy is converted into heat, just as the energy of a stream of water flowing over a waterfall is converted into heat.   The heat produced can easily be calculated.   It is found that to produce 1 calorie of heat (that is to heat 1 gram of water through 1° C.) requires the expenditure of $4 \cdot 2 \times 10^7$ ergs, or $4 \cdot 2$ joules.   The heat produced by a current of C amperes flowing for *t* seconds between points at a difference of

potential of V volts is therefore $\dfrac{1}{4\cdot2}$ × C.t.V calories, or 0·24
C.t.V calories.   Thus

Calories = 0·24 (amperes) × (volts) × (seconds).

In the case of the X-ray tube a very small fraction (less than
1 per cent.) of the energy supplied to the tube is transformed
into X-rays and radiated out from the tube.   The whole of the
mainder is transformed into heat.   The heat generated is,
erefore, very considerable.   In the example given in the
revious section the heat produced would be sufficient to heat
a pint of water to the boiling point in less than seven minutes.
It can be seen that a tube may easily be overheated and spoiled
if it is run for too long at a stretch.

18. ELECTRICAL RESISTANCE ; OHM'S LAW.—It is obvious
that for a given head of liquid more will flow in a given time
through a wide tube than through a narrow one.   Similarly it
is found that for a given difference of potential more current
will be produced in a thick wire than in a thin one of the same
material, and through a short wire than through a long one.
The molecules of the wire impede the free flow of the electrons
through it, and may, therefore, be said to offer resistance to the
current.   The resistance offered by a conductor can be measured
by the difference of potential which must be applied to its ends
in order to produce a given current in it.   A wire which requires
a big difference of potential to produce a given current obviously
has a high resistance.   Thus the resistance of a conductor is
denned as the ratio of the potential difference between the
ends of the conductor to the current produced in it.   A wire in
which an applied potential difference of 1 volt produces a
current of 1 ampere is said to have unit resistance.   The unit
of resistance is known as the ohm.

It is found experimentally that for a given conductor the
ratio of the applied potential difference, or electromotive force,
as it is often called, to the current produced is constant.   Thus
if E is the electromotive force in volts, C the current in amperes,

$$E/C = R,$$

where R is a constant for the conductor and is known as its

resistance in ohms.   This important result is known as Ohm's law.   It may also be put in the form

$$C = \frac{E}{R}$$

and

$$E = CR,$$

both of which are useful in practice.

The resistance of a wire is directly proportional to its length, and inversely proportional to its area of cross section.   It also depends on the material of which it is made.   Silver is the best of all conductors, but is too expensive for general use.   Copper, which is also a very good conductor, is generally employed. Alloys, such as German silver, manganin, and so on, have a much higher resistance than pure metals.

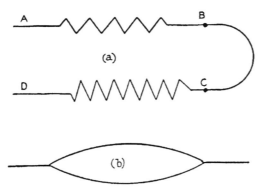

(a)

(b)

FIG. 12.—Conductors in series and parallel.

Solutions of various salts and acids in water conduct appreciably, though their resistance is much higher than that of the metals.   Non-metals, with the exception of carbon, do not conduct appreciably under ordinary circumstances.

Substances which possess the property of resisting the passage of an electric current to a very high degree are called insulators. Rubber, gutta-percha, porcelain, and ebonite are the insulators most generally used in electrical construction.   Thus the copper wires used for conveying the current are generally enclosed in rubber or gutta-percha, which may be further protected by cotton or silk thread, to prevent loss of current by accident: contact with other wires or conductors (short-circuiting, as it is called), and also to safeguard against the dangerous shocks which might be received by any one who should happen to come in contact with them.   Porcelain insulators are used to support bare wires, as in the case of the overhead telegraph wires, while ebonite is generally used to form the handles of electric switches and in the construction of electrical apparatus.

The human body, partly owing to the salts which it contains in solution, is a conductor, though its resistance is high.   Wood, if dry, insulates fairly well, and cork linoleum even better. They thus serve to minimise the danger of shocks produced by accidental contact with high potential wires.   A stone floor is a comparatively good conductor, and is therefore dangerous for electrical departments.

19. SERIES AND PARALLEL CONNECTIONS.—If a number of wires of pieces of apparatus are connected together end to end, so that the current flows through each in turn, they are said to

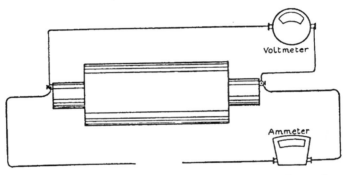

FIG. 13.—Induction coil with ammeter in series and voltmeter in parallel.

be connected *in series*.   Thus the wires AB, BC, CD, in Fig. 12, *a*, are said to be connected in series.   The resistance of a number of conductors connected in series is equal to the sum of their separate resistances.

If two or more wires are joined together at both ends, as in Fig. 12, *b*, they are said to be connected together *in parallel* or in *multiple arc*.   The current on reaching the junction of the two wires has two possible paths, and divides itself so that the larger current flows along the path of least resistance.   The combined resistance of the two conductors in parallel is always less than that of either of the conductors alone.

An ammeter is always connected in series with the circuit through which the current is to be measured.   A voltmeter is connected in parallel with the apparatus.   Thus to measure the current through an induction coil (Fig. 13), we should place the ammeter in series with it. To measure the potential difference between its ends, the voltmeter would be connected in parallel with the coil as shown in the diagram.

20. REGULATION OF THE CURRENT.—It is often necessary to be able to regulate the amount of current flowing through our apparatus. This can be done by changing the voltage of the supply. By the equations in the previous section it can be seen that if we halve the voltage we shall also halve the current; if we double the voltage the current will be doubled. When the current is supplied by a small generator on the pre-

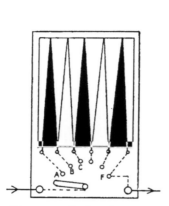

FIG. 14.—Series rheostat.

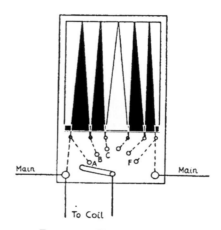

FIG. 15.—Shunt rheostat.

mises arrangements are sometimes made for changing the voltage supplied by the generator. Similarly, if we are using accumulators, we can alter the voltage by altering the number of cells in use. If a battery consists of a number of cells arranged in series, the electromotive force of the battery will be the sum of the electromotive forces of the separate cells.

In general, however, the most convenient way of reducing the current is by increasing the resistance of the circuit. If a length of wire of fairly high resistance is inserted in series with the rest of the circuit, the total resistance will be increased, and thus, by Ohm's law, the strength of the current will be diminished. If the current is to be varied gradually, it will be necessary to be able to include greater or shorter lengths of the resistance wire in series with the circuit. Any device for doing this is known as a rheostat.

The usual form of rheostat is shown in Fig. 14. A number of coils of resistance wire (manganin or other high resistance alloy) are connected to a series of brass studs A, B, C—as shown in the figure. The studs themselves are arranged in a semi-circle, and a stout metal strip can be turned so as to make contact with any one of the studs. The metal strip has insulating handle. The main circuit is broken and one ᵈ of it is connected to the movable strip, the other to the stud F of the rheostat. If the handle is turned so as to ᵏᵉ contact with the first stud A of the rheostat, the current ias to pass in turn through each of the resistance coils. The whole resistance of the wires is thus placed in series with the circuit, and the current is therefore small. If the handle is moved on to B, the coil AB is cut out of the circuit, and the resistance is therefore decreased by the resistance of the coil AB. Thus, as the handle moves from A to F, the resistance in the circuit gradually decreases, and the current consequently increases until it reaches its maximum at F, where all the wires are cut out of the circuit. A rheostat used in this way forms what is known as a series-resistance.

The rheostat may, however, be connected in a different way. The current through a given piece of apparatus, say, an induction coil, can be decreased by allowing some of the current to flow through an alternative path, for example, through a wire connected in parallel with the apparatus. The smaller the resistance of this alternative path the greater will be the current through it, and hence the less the current through the coil. To effect this, the two end studs of the rheostat are connected to the main terminals of the supply; the wires from the coil or other apparatus in which the current is to be regulated are connected the one to the movable handle of the rheostat and the other to one of the end studs, as shown in Fig. 15. If the handle is turned to this stud it will be seen that the induction coil circuit is short-circuited, so that no current will flow through it. As the handle is turned towards the other end of the rheostat, more and more resistance is included in the alternative path of the current, so that more and more of the current flows through the induction coil. When connected in this way the rheostat forms what is known as a shunt resistance.

Either way of using the rheostat will effectively control the current. For various reasons, which cannot be discussed here, it is found that the " shunt " method gives better results in the case of an induction coil, and is therefore generally used in fitting up the switchboard for an X-ray installation. A more elaborate installation may include both forms of regulating the current. The connections for this are shown in Fig. 16.

In the case of the series rheostat an insulated, or " dead," stud is generally provided beyond the stud A, so that when the

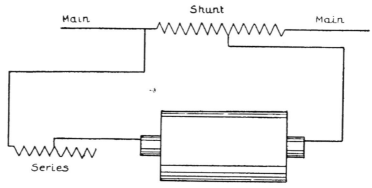

FIG. 16.—Regulation of a current by shunt and series rheostats.

handle is turned on this stud the current is broken and the current stopped. In the case of the shunt method, the current flows through the shunt resistance the whole time until it is cut off by the main switch. It is customary to have a " pilot " lamp on the switchboard connected so that it lights up whenever the main switch is on, to serve as a gentle reminder that the current is running to waste.

Very considerable heat is produced in the wires of the rheostat. They are, therefore, always made of fairly thick bare wire, which, for convenience, is wound into open spirals. Care should be taken that the separate turns of the spiral do not come into contact with each other, and that the wires of the rheostat are at a safe distance from any inflammable material.

21. CONDUCTION THROUGH LIQUIDS.—The conduction of

electricity through metals is, as we have seen, due to the motion of the negative electrons through them, and is not accompanied by any change in the material of the conductor. In the case of the conduction of electricity through solutions of salts in water, the current is carried by the atoms of the elements in the salt, and the salt itself is decomposed in the process.

When common salt, sodium chloride, is dissolved in water we are compelled by many circumstances to believe that the salt is dissociated or split up into its elements. Thus, instead of

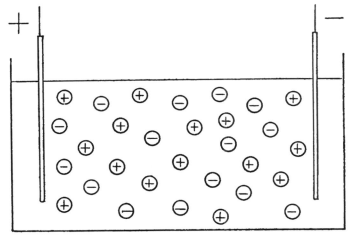

FIG. 17.—Theory of electrolysis.

molecules of sodium chloride, we have in solution electrically-charged atoms of sodium and of chlorine, together with molecules of the undissociated salt if the solution is strong (Fig. 17). The sodium atoms, or ions, as they are called, carry a positive charge, while the chlorine ions carry a negative charge. We can regard this as being due to the transference of one electron from the sodium atom to the chlorine atom which was its partner in the molecule of sodium chloride. If now we place two plates, or electrodes, at a difference of potential in the solution, the positively charged sodium atoms will move towards the negative plate, and the negatively-charged chlorine ions to the positive plate. There is thus a transference of electricity from the one plate to the other across the liquid. In

other words, there is an electric current through the liquid and
the liquid conducts.

On reaching the electrodes the ions give up their charges, and
are liberated as ordinary atoms of the substance. Thus sodium
appears at the negative electrode, or cathode, and chlorine at
the positive plate or anode. If the substance liberated does
not act chemically either upon the water or upon the plate it
will make its appearance in a free state. If chemical action
takes place, only the products of the action will appear. Thus,
in the case of sodium chloride, the sodium reacts with the
water forming sodium hydroxide and hydrogen, which comes
off in bubbles from the electrode. The chlorine is liberated in
a free state, unless the anode is made of a metal which is
attacked by chlorine.

The quantity of a given substance liberated is directly
proportional to the quantity of electricity which has passed
through the solution, that is, to the product of the current and
the time for which it flows. It also depends on the nature of
the substance. Thus, if a current C flows for a time t, the mass
M of a substance deposited is given by

$$M = e \cdot C \cdot t,$$

where e is called the electrochemical equivalent of the substance. The electrochemical equivalent of an element or radicle
is directly proportional to its chemical equivalent. One
coulomb of electricity, that is, 1 ampere flowing for one second,
liberates ·0000104 gramme of hydrogen, twenty-three times
this weight of sodium, or ($\frac{1}{2} \times 63 \cdot 5$) times this weight of copper
(the equivalent of copper being half its atomic weight).

Hydrogen and the metallic elements are all liberated at the
cathode ; non-metals and acid radicles at the anode. The
former are, therefore, positively charged, the latter negatively
charged.

The magnitude of the currents which can be passed through
an electrolytic solution is due to the very large number of
carriers available rather than to the speed with which they
travel. As a matter of fact, the actual velocity of the charged
ions is extremely small, and the distance moved by a given

ion in, say, half an hour is almost negligible. The velocity of the ions is proportional to the electric field acting upon them, that is to say, to the difference of potential between the electrodes divided by their distance apart. The velocity of an ion in a field of one volt per centimetre is known as the mobility of the ion, and depends on the chemical nature of the ion. The mobility of the hydrogen ion is 0·00325 cm. per second, and that of the majority of other ions is only about one-fifth of that of hydrogen. The mobility of the iodine ion, for example, is only 0·00068 cm. per second. Thus, in the case of a current passing between two electrodes 6 cm. apart, and charged to a difference of potential of 12 volts, the individual iodine ions in a solution of potassium iodide between the electrodes would only travel a distance of about 8 mm. in ten minutes. These facts are worth bearing in mind in connection with the theory of ionic medication.

22. ACCUMULATORS, OR STORAGE CELLS.—We can now understand the formation of hydrogen on the copper plate of a voltaic cell. The current flows outside the cell from the copper to the zinc plate. As the circuit must be a complete one, it follows that the current flows through the acid inside the cell from the zinc plate to the copper. The hydrogen ions contained in the solution of the acid, being positively charged, move with the current, and are thus deposited on the copper plate. It is, in fact, the charges given to the copper plate by these ions which give it its positive charge.

The production of this layer of hydrogen on the positive plate is called polarisation. In the case of a " primary " voltaic cell, polarisation is a nuisance which has to be overcome. The effect, however, is made to serve a very useful purpose in the construction of accumulators or storage cells.

If two platinum electrodes are immersed in dilute acid and a current passed for a little while, hydrogen collects on the cathode, and oxygen on the anode. The cell, in other words, becomes polarised. If the current is stopped, and the two plates are connected by a wire, it will be found that a current flows from the oxygen to the hydrogen through the wire, and therefore from the hydrogen to the oxygen through the acid, that is to say, in the opposite direction to that in which the current was originally passed. The arrangement thus forms a little voltaic cell, which continues to act until the gases have all recombined.

This cell differs from the ordinary voltaic cell in the fact that the substances which produce the current have been formed within the cell by the action of an electric current. By passing a current through such an arrangement we put it into a condition in which it can produce current on its own account at some later time. The effect is the same as if the current were stored in the cell (hence the name storage cell), though, as a matter of fact, the energy is stored in a chemical form. Where electrical current can be obtained cheaply from dynamos, accumulators form a very convenient and economical portable source of electric current.

The simple cell described is inconvenient in practice and its capacity is small. The usual type of accumulator consists of a pair of lead grids, which are packed with a mixture of red lead and sulphuric acid. These plates are placed in sulphuric acid (specific gravity 1·18), and a current is passed for some hours. The hydrogen liberated on the one plate reduces the mixture to metallic lead, while the oxygen liberated on the opposite plate oxidises the mixture to lead peroxide. The cell is now charged. If the poles are connected by a wire a current flows from the peroxidised plate to the lead plate, through the wire, the peroxide being gradually reduced to lead monoxide, and the lead plate oxidised to the same compound. When the process has gone so far that the voltage begins to fall, the cell can be restored to its original condition by charging it again from the mains.

These cells work excellently with a little care in handling. Their voltage is remarkably constant (about 2·1 volts), and their resistance is very small. The capacity of the cell, that is the quantity of electricity it can restore, is proportional to the area of the plates. If a large capacity is required a series of plates, alternately positive and negative, may be used, all the plates of each set being connected together to form one large plate. The capacity is stated in ampere-hours, that is to say, the number of hours for which the cell will give a current of

1 ampere after it has been fully charged. The capacity of these cells is quite large. Accumulators made to fit an ordinary pocket flash-lamp will give a current of half an ampere for ten hours, and have thus a capacity of 5 ampere-hours. The larger batteries used for motor-car ignition have generally a capacity of about 40 ampere-hours. A battery of these cells is often used for X-ray work, where facilities exist for having them recharged, but where the electric supply is not actually laid on in the building. They are also, at present, used for the heating circuit of the Coolidge tube.

In using them care should be taken to maintain the acid at the proper strength (specific gravity 1·18), and to have them recharged as soon as the voltage falls appreciably below 2 volts per cell. The cells may be left for a month or more without charging if they have been fully charged, but they should never be allowed to stand in an uncharged or partially charged condition. It must be remembered that it is not possible to take more current from the cell than is put into it, in fact only about 80 per cent. is actually returned. Thus a cell which has a capacity of, say, 40 ampere-hours will require to be charged with a current of 4 amperes for at least ten hours, and preferably for twelve hours, in order to recharge it to its full capacity. There is no danger in charging the cell for a little longer than this, if the current used for charging does not exceed the strength which the cell will safely carry.

A cell with a capacity of 40 ampere-hours would give a current of 1 ampere for forty hours, 2 amperes for twenty hours, and so on. It should, theoretically, give a current of 40 amperes for one hour. Large currents, however, tend not only to overheat the cells, but also to cause the plates to buckle, and for every cell there is a maximum current, depending on the size of the plates, which cannot be exceeded without risk of injury to the cell. Anything up to one-tenth of the capacity of the cell is quite safe. Thus a cell with a capacity of 40 ampere-hours could safely be used to give currents up to 4 amperes. Unfortunately, owing to its low resistance, the cell will, if allowed, give much larger currents than this, and currents of 10 or 20 amperes could easily be obtained from such a cell.   The tendency to overrun the cells when using them for radiography is a very natural one, especially as an installation running on cells is apt to be rather underpowered, but it should be severely repressed.   The results may not be immediately apparent, but the practice will inevitably lead to a rapid dis-integration of the plates, and consequent expensive renewals. Care should be taken when installing the cells to see that their capacity is ample for the work they are required to do.

Accumulators are charged by connecting the positive pole to the positive terminal of the supply and the negative pole to the negative terminal.   A rheostat and ammeter should be included in the circuit.   The voltage of the supply should be at least 20 per cent. higher than that of the cells to be charged.   The current should be adjusted to the maximum charging current (which is usually stated by the makers), and as it is apt to fall during the first hour or so of the charge, it is advisable to read the ammeter occasionally and adjust the current again if required.   Cells, even if not in use, should be charged at least once a month.

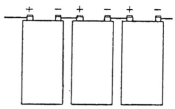
FIG. 18.—Accumulators in series.

If a set of cells is connected so that the positive of one cell is connected to the negative terminal of the next, the cells are said to be connected in series.   The voltage of the battery is then the sum of the voltages of the separate cells (Fig. 18). Thus a battery of twenty-four accumulators, each having an electromotive force of 2·1 volts, will have a difference of poten-tial of 24 × 2·1, that is 50·4 volts between its extreme ends.   If two cells are taken and their two positive plates are connected together, and also their two negative plates, they are said to be connected in parallel.   The potential difference will be that of a single cell, but, owing to the fact that only half the total current supplied by the battery comes from each cell, they can be used to give twice the current which could safely be taken from a single cell.   The parallel arrangement can also be used in charging the cells, if the voltage of the supply is not sufficient to charge the whole of the cells in series.   Thus a battery of twenty-four cells could be charged from a supply at 30 volts by dividing them into two sets of twelve in series, and connecting the sets in parallel, as shown in Fig. 19.   As each set would only take half the whole current it would, of course, be necessary to employ twice the normal charging current, or else to charge the cells for twice the usual time.   Direct current must of course be used for charging the cells.

We have made this digression as accumulators are increasingly being used in Radiography, and it is much more convenient, and much safer, to charge them on the spot, if at all possible, than to trust them to the tender mercies of a local garage.   If direct current is installed in the department, a proper charging board, with ammeter and rheostat, should certainly be installed for charging the cells.   The directions which the makers usually attach to the cells should always be rigidly adhered

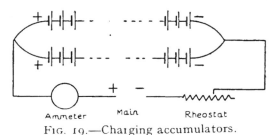

Ammeter   Main   Rheostat

FIG. 19.—Charging accumulators.

to.   The acid should be tested from time to time by an hydrometer (small hydrometers specially made for the purpose can be procured), to see that it has the right specific gravity (1·18 after charging).   If the specific gravity is too high, it can be adjusted by adding a little distilled water.   If too low, a little pure sulphuric acid may be added.   The acid in the battery should always be just sufficient to cover the plates.

Accumulators cannot, of course, be charged directly from the alternating current mains, as alternating current is always reversing its direction.   Rectifiers of various kinds can, however, now be obtained which allow the current to pass through them only in one definite direction.   If a rectifier is placed in series with the alternating current, the latter thus becomes unidirectional, and can be used for charging the accumulators. The older types of rectifiers were messy, and not very efficient. In the latest patterns, however, these defects seem to have been overcome, and they provide a very handy means of charging accumulators where only alternating current is available.

CHAPTER III

THE INDUCTION OF CURRENTS

23. THE INDUCTION OF CURRENTS.—If two coils of wire, A and B (Fig. 20), are placed parallel to each other at a short distance apart and a current is started in one of the coils, it is found that a momentary current flows in the other coil. This phenomenon, which is known as electromagnetic induction, was discovered by Faraday, and is the starting point of nearly all the technical applications of electricity.

Thus if the coil A (Fig. 20), which we will call the primary coil, is connected to a battery and a tapping key K, while the

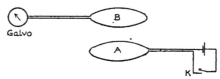

other, B, is connected in series with a sensitive galvanometer for detecting small currents, on allowing the current to flow in A, by depressing the key, the galvanometer in the B circuit (which is known as the secondary circuit) gives a sudden deflection, showing that a current is also flowing in B. This current is quite transient, and only flows as long as the current in A is varying. As soon as the latter becomes steady, which takes only a minute fraction of a second as a rule, the induced current in B stops.

FIG. 20.—The induction of currents.

If now the current in A is stopped by raising the key and so breaking the circuit, a transient current again flows in B, but in the opposite direction. These transient currents are known as induced currents. The current in B when the circuit in A is made flows round the coil in the opposite direction to that of the current in A, and is known as the inverse current. The current produced in B when the primary circuit is broken is in the same direction as the primary current in A, and is known as the direct current.

The induced current in B is not a direct effect of the current in A, but is due to the fact that the latter produces a magnetic field of force in its neighbourhood, some of the lines of which will pass through the coil B. It is the introduction of these lines of force through the secondary circuit which causes the inverse induced current to flow, while their withdrawal, when the current in A is stopped, produces the direct induced current. Similar effects can be produced by using a magnet. Thus if one pole of a bar magnet is brought near the coil B, a current will flow in the latter while the magnet is approaching the circuit, and a current in the opposite direction while the magnet is being withdrawn.

It is found that the quantity of electricity flowing round the secondary circuit is directly proportional to the change in the number of lines of magnetic force passing through the circuit. If, however, the circuit consists of, say, ten turns of wire, each line of force will pass through the circuit ten times. The effect is thus proportional to the number of turns of wire on the coil. The quantity of electricity passing round B is also inversely proportional to the resistance of the circuit.

Thus, to return to the case of the two circuits, when a current is passed through A a certain number of its magnetic lines of force will pass through the coil B, and a quantity of electricity will be sent through the galvanometer equal to the number of lines of force which pass through B divided by the resistance of the secondary circuit. The effect is therefore directly proportional to the strength of the primary current, and increases as B is brought nearer and nearer to A. If the coil B is placed so that it completely surrounds A, then the whole of the lines of magnetic force produced by the primary circuit pass through the secondary, and the effect is a maximum.

24. INDUCED ELECTROMOTIVE FORCE.—The quantity of electricity set in motion depends only on the total change in the number of lines of magnetic force, and is independent of the speed with which the alteration is brought about. The induced electromotive force in the secondary circuit which produces the secondary current is, on the contrary, directly proportional to the rate at which the lines of force are chang-

ing. If the change is made slowly the induced E.M.F. in the secondary will be small. If the change is made very rapidly the induced E.M.F. will be high. Thus, if the current in A is gradually increased up to its maximum value, the quantity of electricity flowing round the secondary circuit B will be the same as if the current were at once switched on at its maximum value, but the induced E.M.F. will be much smaller.

The induced current can be largely increased by winding both the primary and secondary coils upon a single core of iron. The iron has the effect of greatly increasing the number of lines of magnetic force produced by a given current, and hence the induced secondary current is largely increased, often more than one thousand times. In one of his early experiments

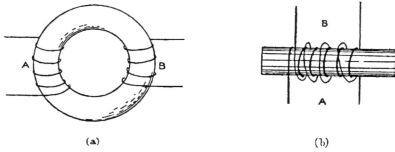

(a)   (b)

FIG. 21.—Faraday's induction experiments.

Faraday took a circular ring of iron and wound two coils upon it, as shown in Fig. 21, *a*. On passing a current through A, a very strong inverse current was produced in B. This ring of Faraday's is the prototype of all transformers. In another experiment the two coils were wound, one inside the other, on a straight bar of iron (Fig. 21, *b*). The effects were similar to those obtained with the ring. This arrangement is effectively that of the induction coil.

Since the E.M.F. induced in the secondary is proportional to the rate at which the lines of magnetic force are withdrawn from it, it is obvious that by making the changes sufficiently rapidly we can obtain very high potential differences in our secondary circuit. If there are n turns of wire on the secondary coil, the withdrawal of a single line of force from the primary circuit will effectively remove n lines of force from the secondary, since each line of force cuts the secondary circuit n times. Thus, to obtain a high induced E.M.F., we require a large number of turns of wire on the secondary coil. In the case of the transformer it can be shown that the ratio of the induced E.M.F. in the secondary to the E.M.F. in the primary is approximately equal to the ratio of the number of turns of wire on the secondary to the number of turns on the primary. Thus, if the primary coil consists of 10 turns of wire and the secondary of 10,000, the secondary E.M.F. will be 1,000 times that in the primary. Thus, by increasing the ratio of the number of windings, current at a comparatively low E.M.F. can be transformed into current at a very high E.M.F. It is this fact which gives induction coils and transformers their importance in Radiology.

It is obvious that we cannot get more power out of the secondary of the coil than we put into the primary. Thus, when the secondary voltage is high, the strength of the current will be comparatively small. If, for example, the primary coil is taking an average current of 5 amperes at 100 volts, the power supplied to the coil is $5 \times 100 = 500$ watts. The power in the secondary circuit cannot exceed this, and so, if the secondary voltage is 100,000 volts, the current cannot be greater than $\dfrac{5}{1000}$ ampere or 5 milliamperes. Owing to various inevitable losses of energy in the system, it would, in practice, be less than this, probably about 3 or 4 milliamperes if the coil were properly constructed, and less if it were badly designed.

It must be noticed that the secondary or induced current only flows while the primary current is changing in strength. Hence, to use a transformer or an induction coil as a source of current, we must arrange that the primary current shall be always changing in strength. In the case of transformers this is usually done by supplying the primary with what is known as alternating current.

25. ALTERNATING CURRENTS.—Currents are called continuous or direct if they flow, without stopping, always in the same direction. They are called alternating currents if they are continually reversing their direction in a regular periodic

manner, so that they flow first in one direction and then in the opposite direction round the circuit. An alternating current, therefore, starts at zero, rises to a maximum value in one direction, then decreases to zero, increases again, but in the opposite direction, and so back to zero ; the cycle being continually repeated as long as the current is flowing. The number of complete cycles made per second is called the frequency of the current. The alternating current of the electric supply generally has a frequency of from 50 to 150 cycles per second.

Alternating current is produced by special alternating current dynamos. Alternating currents are themselves induced currents produced by rotating a coil of wire in a strong magnetic field. If a coil of wire is rotated uniformly in the strong magnetic field between the poles of a horse-shoe electromagnet, the number of lines of magnetic force passing through the coil is constantly varying, being a maximum when the plane of the coil is at right angles to the field, and zero when the plane is parallel to this direction. There will thus be an induced E.M.F. in the coil which will be proportional to the rate at which it cuts the lines of force. Moreover, as the direction of the lines of force relative to the coil changes every half revolution, the direction of the induced E.M.F. will also change every half revolution, and if the two ends of the coil are connected to an electric circuit, a current will flow first in one direction and then in the other through the circuit, changing in direction at each half revolution of the coil. This is the principle of the alternating current dynamo in its simplest form. The electric current supplied by the supply companies is generally alternating, owing to the fact that alternating current is cheaper to transmit over long distances than direct current, while being just as convenient for lighting and heating purposes.

26. PRINCIPLE OF THE TRANSFORMER.—If we connect the primary of a transformer with the alternating supply, then, as the current is always changing its magnitude and direction, induced current will be continually flowing in the secondary coil. As the primary current changes its direction the secondary current will also change in direction. The secondary induced

current will thus also be alternating current having the same frequency as the primary. Thus the transformer can be used to transform a low potential alternating current into a high potential alternating current in the simplest possible way.

Unfortunately, it is most important for radiographical purposes that our high tension current shall be direct, and flow only in one direction round the circuit. This can be effected by using a motor which rotates in exactly the same frequency as the current, and which is arranged so that it breaks the secondary circuit during the time when the current would be flowing in one direction, but makes it when the current is in the opposite direction. The current through the secondary circuit then consists of a series of pulses, or rushes, of current all in the same direction. This is the principle of the high tension transformers, such as the Snook. The device works quite efficiently, but of course it considerably increases the complexity of the apparatus.

27. PRINCIPLE OF THE INDUCTION COIL.—In the case of the induction coil direct current is always used, and some mechanical device, known as an interruptor, is placed in the primary circuit which continually makes and breaks the circuit many times per second. These are described in a later section (§ 44). When the current is made an inverse current flows in the secondary, when the current is broken a direct current flows. The current in the secondary is therefore alternating. The peculiarity of the induction coil, or open circuit transformer, as it is called, is that, by proper adjustments, the current at make can be practically suppressed, so that only the direct current produced by breaking the primary circuit is effective.

We have seen that the introduction of lines of magnetic force through a circuit produces an induced E.M.F. in the circuit. Now, when a current is started in the primary coil it will produce lines of force, not only through the secondary, but also through its own circuit. There will thus be an inverse or " back " E.M.F. induced in the primary circuit when the current is made, which, as it acts in the opposite direction to the current, will tend to retard its rise in strength. This effect is known as the self-induction of the circuit, and its value can be ascertained both by calculation and by direct experiment. Owing to self-induction the current, on closing the key, does not immediately rise to its full value, but takes a certain amount of time to reach full strength, which is proportional to the ratio of the self-induction to the resistance of the circuit. If the resistance is small and the self-induction large, the rise of the current will be comparatively slow. On the other hand, if the resistance is very large or the self-induction small, the current will reach its steady value very rapidly.

Now suppose that the primary of the induction coil has a fairly high induction and a low resistance, as is always the case, on making the circuit the current will rise comparatively slowly to its final steady value, and as the induced E.M.F. in the secondary is proportional to the rate of change, the induced E.M.F. at make will be comparatively small. When, however, the primary current is broken the conditions are altered. By breaking the circuit we introduce into it an air gap of almost infinite resistance. But the time taken for the current to fall to its final value (in this case zero) is inversely proportional to the resistance of the circuit. The current, therefore, on breaking the circuit falls with great rapidity, and the induced E.M.F. in the secondary is correspondingly high. Thus, the direct E.M.F. in the secondary at break is much higher than the inverse E.M.F. in the secondary at make.

If the secondary circuit were closed by a conducting wire, the quantity of electricity flowing round the secondary circuit at make and at break would be exactly equal. This is due to the fact that the secondary E.M.F. at make, though much smaller than that at break, acts for a much longer interval of time. The induced E.M.F. at make is comparatively feeble, but prolonged ; that at break is intense, but acts for a very short time. As a matter of fact, however, in the class of work for which an induction coil is used the secondary terminals of the coil are never joined by a conductor. They are brought to the two poles of a spark gap of some kind. Now we have seen that air has the peculiarity of being an almost perfect insulator for voltages of less than some critical value (the sparking potential for the distance between the poles of the gap). If, however, the sparking potential is exceeded, the resistance of the air breaks down and a considerable current may pass.

If the coil has been properly designed there will be a considerable range of spark gap for which the inverse induced E.M.F. at make will be too small to spark across the gap, though the direct induced E.M.F. at break will be sufficiently high to do so. The actual current through the secondary will in such cases always be in the same direction, namely, direct. The inverse current will thus be completely suppressed. We shall see later that an X-ray tube resembles a spark gap in requiring a definite minimum voltage to excite it. The current produced through an X-ray tube by a well-designed coil should therefore be unidirectional.

28. THE EFFECT OF THE CONDENSER ON A COIL.—It is found that when a mechanical make and break is used, the working of the induction coil is enormously improved by connecting a condenser of suitable capacity across the points in the primary circuit between which the interruptions take place. The effect is not easy to explain in an elementary manner; in fact, it is doubtful whether the action of the condenser is completely understood. The very rapid fall of the primary current at break induces a considerable E.M.F. not only in the secondary, but also in the primary itself, since lines of force are being rapidly withdrawn from the primary circuit. This induced E.M.F., which is, of course, in the direction of the primary current, tends to spark across the terminals of the interruptor. This tends to prolong the primary current, and thus to impair the rapidity of the break upon which the efficiency of the coil depends. If, however, there is a condenser in parallel with the gap, the extra current, instead of sparking across the gap, flows into the condenser, and the actual break is thus made more sudden. As the primary E.M.F. falls still further the condenser discharges itself round the primary circuit in the opposite direction to the primary current. The condenser discharge thus helps the rapidity of the fall of the last portions of the primary current, and also assists in reducing

any remaining magnetism in the iron core of the coil. In brief, the function of the condenser is to reduce the sparking at the contacts of the break, and thus increase the suddenness of break upon which the action of the coil depends. Condensers are not employed with electrolytic interruptors.

The effect of the condenser is probably more complicated than is suggested by this explanation, and a careful adjustment between the capacity of the condenser and the self-induction of the primary is required to obtain the best results. The theory of the induction coil is still far from having been completely worked out. Further research may result in a considerable improvement in the design of induction coils for X-ray work.

CHAPTER IV

29. CONDUCTION OF ELECTRICITY THROUGH GASES.—Gases in their normal condition are among our best insulators. They can, however, be rendered partially conducting either by raising them to a high temperature, or by the action of various radiations, known as ionising radiations, of which X-rays and the various radiations from radium are the most important.

If an X-ray tube is worked near a charged electroscope the latter rapidly loses its charge, showing that the surrounding air has lost its insulating properties. The loss is only temporary. If the rays are stopped the air ceases to conduct.

It has been shown that the conductivity is due to the formation within the gas of positively and negatively charged molecules, which, from analogy with the phenomena of electrolysis, are known as ions. X-rays exert an electric force on the electrons, which, as we have seen, are contained in all atoms, and in favourable circumstances may cause one of them to be ejected from the atom, leaving it positively charged. The negative electron thus set free attracts an uncharged molecule in its neighbourhood and, sticking to it, gives it a negative charge. Thus the gas, which originally consisted only of neutral molecules, now contains molecules which are positively and negatively charged. If a positively charged body is now placed in the gas it will attract the negative ions, and those which reach it will give up their charges to it, and thus neutralise a portion of its charge. Similarly, a negative body would lose its charge by attracting the positive ions. If a pair of plates at a difference of potential are immersed in the gas, a current will flow from the one to the other, in much the same way that the current is conducted across an electrolyte. The current is always small, generally far too small to be detected by any of the instruments generally used for detecting currents. It can, however, be measured by the more sensitive method of finding the rate at which an electroscope of known capacity loses its charge.

The gaseous ions differ from the electrolytic ions in the fact that they are not formed spontaneously, but are manufactured by the action of some external agent. The number formed per second is proportional to the strength of the ionising radiation. When the latter is cut off no fresh ions are formed, while those which are present, owing to their being oppositely charged, attract each other. On meeting, their charges neutralise each other, so that they part as ordinary uncharged molecules. The conductivity of the gas, therefore, rapidly ceases when the ionising agent is removed.

This recombination of the ions takes place even while the rays are acting upon the gas. If, however, the electric force applied to the gas is sufficiently strong, every ion will be attracted to one or other of the electrodes so quickly that it will reach it before it has had time to collide or recombine with an ion of opposite sign. Under these circumstances all the ions formed in the gas will give up their charges to the appropriate electrode, and the charge gained by the electrodes will be proportional to the number of ions formed in the gas. As the number of ions formed per second is proportional to the strength of the ionising agent, the rate of loss of charge of either of the electrodes is a measure of the intensity of the ionising agent. This is the most accurate way we have, at present, of measuring the intensity of a beam of X-rays.

30. MEASUREMENT OF IONISATION CURRENTS.—The measurement can be performed using an ordinary gold-leaf electroscope of the type figured in Fig. 1, if one of the sides is made of thin aluminium so as to be transparent to the rays. Let us suppose that the gold-leaf is charged to some potential V, as recorded on the scale of the instrument, and the rays are then allowed to enter it. All the ions of the opposite sign to the leaf will be attracted to the insulated system, the others being driven to the case which is earthed. The gold-leaf system thus begins to lose its charge and the potential falls.

Let t be the time taken to fall to some lower potential V'. This time can be measured with a stop-watch.

Now if C is the capacity of the gold-leaf system the loss of charge Q which has taken place in the time t is given by

$$Q = CV - CV' = C(V - V'),$$

and this is equal to the charge conveyed to it by the ions in the given time. Now the number of ions formed in this time is proportional to the intensity of the rays, and to the time t during which they have been acting. As every ion has the same charge, the total charge conveyed to the electrode is therefore kIt, where I is the intensity of the rays, and k a constant which will depend on the size of the electroscope, and also on the quality of the rays. Thus we have

$$kIt = C (V - V')$$
$$or\ I = \frac{C (V - V')}{kt}$$

For the same electroscope and the same kind of rays, both C and k will be constant, and by always taking the time required by the leaf to fall between the same two graduations on the scale, V -- V' can also be made constant. In this case the intensity of the rays will be inversely proportional to the time taken for the leaf to fall from one definite position to another. Thus if, on one occasion when the rays were passed into the electroscope, the leaf was found to fall from the fortieth to the thirtieth division of the scale in twenty seconds, while on another occasion it took sixty seconds to pass between the same two divisions, the intensity of the rays in the first case would be three times the intensity in the second.

31. APPLICATION TO THE MEASUREMENT OF RADIATION.— The total charge lost by the charged system is obviously proportional to the total quantity of radiation which has entered the electroscope. Thus for rays of the same quality the fact that the gold leaf has fallen from the fortieth to the thirtieth division implies that a definite quantity of radiation has entered the electroscope. The instrument can therefore be used as a quantimeter. Instead of passing the rays directly into the electroscope it is often more convenient, and more accurate, to

4—2

pass them through a small metal box containing an insulated electrode which can be connected by a wire with the electroscope or other measuring instrument. A convenient form is shown in Fig. 22. It consists of a flat, cylindrical metal box, the two ends of which are closed by thin aluminium foil, so as to be transparent to the rays. The inner electrode consists of a circular plate of thin aluminium placed parallel to the ends of the box, and insulated from it by ebonite or amber. The inner electrode is connected to the gold-leaf system of the electroscope, and the case to earth. The gold leaf and the connected electrode are charged, and the rays are passed through the ionisation chamber as it is called, rendering the gas there conducting. The method of making the measurements and the principles involved are exactly the same as if the rays were passed directly into the electroscope.

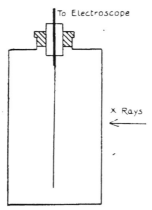

To Electroscope

x Rays

FIG. 22.- Simple ionisation chamber.

An instrument on this principle, using a multicellular voltmeter (§ 10) instead of a gold-leaf electroscope, has been placed on the market under the name of the iontoquantimeter. It suffers from the defects of all quantimeters, that its readings depend not only on the quantity but also on the quality of the rays. On the other hand it gives definite numerical results, and is capable of greater accuracy than the coloration methods generally used in radiographic departments. It is the only method employed in scientific work on the subject, and, given apparatus of suitable design, there is no reason why it should not be made as convenient in practice as the less accurate methods now in use.

The method has been employed for a considerable time for measuring the intensity of the radiations from radium and other radioactive substances. Quantities of radium are now always measured by comparing the intensities of the radiation from the sample to be measured at some standard distance with that from a radium standard, of known quantity, at the same distance. For this purpose the simple electroscope method (without an ionisation chamber) is generally preferred.

32. THE ELECTRIC SPARK.—We have already seen that, while air under ordinary conditions is a good insulator, yet, if the electric force rises sufficiently high, the insulation breaks down and a spark passes. Thus if the two ends of the secondary of an induction coil are connected to two metal rods, and the latter are gradually approached, at a certain distance, depending on the induced electromotive force in the secondary, a spark will pass between them. The passage of this spark renders the air temporarily a comparatively good conductor,

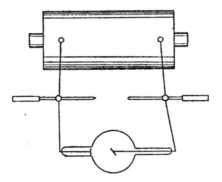

FIG. 23.—Alternative spark gap.

ind a considerable current flows across the gas. When this ırrent ceases owing to the falling off of the electromotive force, the air again becomes non-conducting until the passage of another spark.

A spark gap of some kind is almost invariably fitted to all ıduction coils as a means of estimating the potential difference ʒoduced by the coil. This spark gap is generally fitted with ıng cords, so that the two rods can be moved while the coil is actually working, and a scale on which their distance apart can be read off. It is then sometimes known as a spinktermeter. Unfortunately the relation between the spark length and the potential difference is not a simple one, and depends very largely indeed on the shape of the sparking points. We have already mentioned that a spark passes much more readily between points than between parallel plates, and it may be stated generally that the more pointed the conductors the greater will be the distance at which sparking will take place for a given difference of potential. The voltage given by the secondary of an induction coil is generally stated in terms of the length of spark which it will give. This will obviously be much greater if measured between sharp points than if measured between, say, two spherical conductors. In the same way, the voltage required to excite an X-ray tube is always measured in practice by means of what is known as an alternative spark gap. The spark gap is connected in parallel with the tube (Fig. 23), and the rods are gradually drawn nearer together until sparks begin to pass. If this happens when the ends of the rods are, say, 2 inches apart, the tube is said to have an alternative spark gap of 2 inches, and is described not infrequently as a 2-inch tube. The actual voltage across the tube, when the alternative spark gap is 2 inches, will, however, depend very largely on the shape of the discharging " points " of the spark gap. If these are spherical, with a radius of 2·5 cm., this spark gap will correspond to a potential difference of 84,000 volts ; if they are blunt points, to a voltage of 56,000 volts ; while if the points are very sharp, the voltage may be as little as 13,000 volts. It is obvious that the actual condition of the tube would be very different in the three cases. These facts are apt to be overlooked.

The spark gaps fitted to induction coils are generally made with pointed conductors, or sometimes with one point and one small plate. The use of points is to be deprecated, as the corresponding voltage varies very rapidly with the sharpness of the point. If some agreement could be reached as to a standard shape and size for the discharging " points," it would certainly increase the definiteness of this method of estimating the condition of the tube.

The following table, giving the voltage corresponding to sparks of various lengths, under different conditions, may be interesting in this connection, though too much reliance should not be placed on the actual figures, as the results obtained by different observers are far from being concordant.

TABLE I.

Potential Difference in Volts required to produce a Spark in Air.

Spark length in inches.	Sharp Points.	Dull Points.	Ball Electrodes. 2·5 cm. radius.	Point and Plane.
1	11,000	25,000	65,000	—
2	13,000	30,000	84,000	—
3	—	40,000	90,000	—
4	—	73,000	92,000	110,000
6	—	94,000	95,000	—
8	—	119,000	—	150,000
12	—	165,000	—	190,000
16	—	210,000	—	230,000

CHAPTER V

X-RAYS

33. THE DISCHARGE TUBE.—If two metal terminals are sealed into a glass tube, and the air is gradually exhausted, it is found that the sparks pass more and more readily as the pressure is reduced. As the pressure gets lower the spark also becomes wider and more diffuse until, finally, it fills the whole tube. At this stage it generally breaks up into a series of luminous striæ, or glows, which in air are of a reddish colour. The appearance of this reddish glow in an X-ray tube is generally the fatal sign that the tube is punctured. At this stage the current passes through the discharge tube with considerable ease, a potential of about 400 volts being sufficient to maintain the discharge in air at a pressure of 2 mm. of mercury.

If the evacuation is carried further by means of a good vacuum pump, a dark space, known as the Crookes dark space, is seen forming at the cathode, and as the pressure is still further reduced this dark space grows, driving the various glows before it, until eventually the whole tube may be practically dark, even though the current is still passing. During this stage the potential difference necessary to pass a current through the tube rises rapidly, and may exceed that required when the tube was filled with air at atmospheric pressure. It is, in fact, possible to exhaust a tube so completely that the highest potentials yet produced are insufficient to cause a discharge.

If the tube is closely observed at the stage when the Crookes dark space is well developed, but the pressure is not yet too low, a set of bluish streamers can be seen proceeding normally from the cathode or negative terminal and crossing the dark space. These are known as the cathode rays. The cathode rays consist of negative electrons shot off from the cathode with a very high velocity. It is, indeed, by observations on the cathode rays that most of our knowledge of electrons has been gained. The electrons themselves are not visible, the bluish glow being due to the fluorescence they excite in the residual gas in the tube. This is shown by the fact that if the tube is highly exhausted the cathode beam is not visible though its existence is demonstrated by the greenish fluorescence produced where it strikes the glass walls of the tube. The stream of cathode rays can be seen quite well in an X-ray tube which is rather too "soft" for radiographic purposes (2-inch to 3-inch spark gap). It is not visible in a medium or hard tube.

34. PRODUCTION OF X-RAYS.—It was while working with a discharge tube of this kind that Rontgen, in 1894, made his great discovery. He noticed that when the tube was excited a fluorescent screen, consisting of crystals of barium platino-cyanide, became luminous, even if it were turned so that none of the light from the discharge fell upon its surface. He also found that photographic plates enclosed in light-proof envelopes became fogged if left near the tube, just as if they had been exposed to light. These effects were found to be due to something which was radiated in straight lines from the spot on the walls of the tube where the cathode rays fell. This new radiation Rontgen called X-radiation, to indicate that its nature was unknown. It is only quite recently that the nature of the rays has been definitely determined by Laue, though it had been long suspected. X-rays are now known to be electromagnetic radiations of the same nature as ordinary light, but of very much smaller wave-length, produced by the sudden stoppage of the negatively charged electrons. Their wave-length has been still more recently determined by Bragg, and is of the order of 10^{-8} to 10^{-9} cm. The wave-length of the most luminous part of the light spectrum is about 5×10^{-5} cm.

35. PROPERTIES OF THE RAYS.—X-rays are produced at the spot where the cathode rays strike the target, and radiate outwards from it in all directions in straight lines. Their intensity thus decreases as the distance from the source is increased. Suppose, for example, the rays from a point X

(Fig. 24) pass through a square hole ABCD at a distance of 20 cm. from X. The rays travel on in straight lines, those through the corners of the square being indicated by the lines in the diagram. If a screen S is placed some distance behind the aperture the rays passing through ABCD will illuminate a square patch on the screen, which, as can be seen from the diagram, is considerably larger than the aperture itself. As the same quantity of rays have to illuminate a larger area, it is obvious that their intensity, that is to say, the quantity falling on unit area, will be less. It can easily be shown by simple geometry that the area EFGH is to the area ABCD directly as the squares of their respective distances from X.

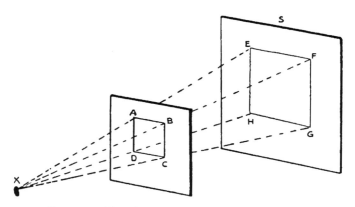

Fig. 24.—Proof of the "inverse square" law.

The intensity of the radiation is, therefore, inversely proportional to the square of the distance from the origin of the rays. Thus, if the distance of S from X is twice the distance of ABCD from X, the intensity of the rays on the screen will be one-quarter of their intensity at ABCD, and they will require four times as long to produce the same effect. In the same way, the intensity of the rays at a distance of 35 cm. from the target will be $\frac{50^2}{35^2}$, or very nearly twice the intensity at a distance of 50 cm. These considerations are of great importance in estimating exposures.

X-rays do not excite the sensation of vision, and surfaces on which they are falling do not, as a rule, appear luminous.

Certain substances, however, such as barium platino-cyanide, zinc blende, and willemite, fluoresce when struck by the rays, and can, therefore, be used for detecting their presence. A fluorescent screen consists of a uniform layer of one of these substances, generally the first, mounted on stiff cardboard. The cardboard is transparent to the rays, so that the latter pass through it to the layer of fluorescent substance, which then shines more or less brightly, according to the intensity the rays. The fluorescent screen is covered with a sheet lead glass, partly to protect the delicate surface from dirt and abrasion, and partly to protect the operator from the direct action of the rays. Lead glass, while transparent to ordinary light, absorbs X-rays to a considerable extent.

X-rays affect a photographic plate in the same way as light. As in the case of light, the image is not visible until the plate has undergone chemical treatment known as development. The parts acted upon by the rays then appear dark, while the parts which have not been acted upon appear as transparent glass. The density of the image depends on the quantity of radiation which has fallen on the plate. The emulsion used on ordinary photographic plates is too thin to stop all the X-rays. A considerable percentage passes through without producing any effect. The density of the image can, therefore, be increased by using a thicker emulsion, and so stopping more of the rays. The special X-ray plates differ from the ordinary plates mainly in the thickness of the emulsion. Their after-treatment is just the same as that for plates exposed to light in an ordinary camera.

On account of their very short wave-length X-rays cannot be either reflected or refracted in the ordinary sense of the terms. They cannot, therefore, be made to give images, in the way in which a photographic lens produces an image of a landscape or other object. X-rays are, however, scattered by material substances, through which they pass in much the same way that a beam of light is scattered in passing through a slightly turbid medium, such as may easily be obtained by pouring a few drops of milk into a tumbler of water. These scattered, or secondary, rays are much less intense than the primary beam which produces them, but the fact that a body through which X-rays are passing becomes itself a source of X-radiation, radiating out from it in all directions, is one which requires to be borne in mind.

36. ABSORPTION OF X-RAYS.—X-rays are able to pass through considerable thicknesses of many substances which are opaque to ordinary light. It is this property which has rendered them so valuable as an aid to diagnosis. No substance is perfectly transparent to X-rays, and no substance is absolutely opaque. In every substance there is a gradual absorption of the X-rays, the amount absorbed increasing with the thickness. In some substances, however, the absorption is comparatively small, so that the rays can pass through very considerable thicknesses of the substance. In other substances the absorption is so rapid that a thickness of a few millimetres is sufficient to cut off practically the whole of the rays. In general, substances of low density and containing only elements of low atomic weight are very transparent to the rays, while dense substances, especially if of high atomic weight, are very opaque. Thus flesh, wood, wool, cotton and paper are very transparent; aluminium and ordinary crown glass are somewhat less transparent. Healthy bone is fairly opaque to the rays, while the majority of the metals, such as iron, copper, nickel and lead, are very opaque, a comparatively small thickness of these substances being sufficient to cut off nearly the whole of the rays.

Since the X-rays travel in straight lines, they will cast shadows of obstacles in their path in just the same way as light. The shadows can be made visible by using a fluorescent screen. If a piece of lead, sufficiently thick to stop practically the whole of the rays falling on it, is held between the source of the rays and a fluorescent screen, no rays will strike the part of the screen within the shadow of the lead, and that part will, therefore, appear dark. If a partly transparent object is substituted for the lead, the portion within the shadow will be less vividly illuminated than those parts without, as a portion of the rays will have been absorbed by the object. The shadow cast will, therefore, not be so dense as in the case of the lead. In other

F<small>IG</small>. 25.—Radiogram of fracture of the humerus. The dark shadows cast by the bones show distinctly through the lighter shadow cast by the flesh.

words, the contrast between the shadow and the rest of the screen will be less.

Thus opaque objects cast dense shadows, while fairly transparent ones cast only faint shadows. Thus in the case of a limb the shadow cast by the bones is much darker than that cast by the flesh, and hence is distinctly visible within the fainter shadow of the flesh (Fig. 25). It should not be forgotten that the appearances are shadows, and not images, of the various objects.

37. CHARACTERISTIC X-RADIATION.—When a beam of X-rays falls upon an element, in addition to the general scattered radiation which is similar in quality to the primary beam, the element also gives out secondary X-rays of a softer type, which are found to be of one or two definite wave-lengths and are characteristic of the element. Just as a sodium salt heated in a Bunsen burner emits the well known D lines of sodium, so a copper plate when acted upon by a beam of X-rays gives off radiation of definite wave-lengths which are as characteristic of copper as the D lines are of sodium. Every element has thus its characteristic X-ray spectrum. This characteristic radiation is only excited if the incident rays are of shorter wave-length than the characteristic radiation of the element. If the incident beam is of greater wave-length than the characteristic rays only the scattered radiation appears.

If the primary rays are of suitable wave length the characteristic rays from an element are found to consist of two kinds, one of which is much more penetrating, *i.e.*, of much shorter wave-length, than the other. The more penetrating of the two is called the K radiation, the less penetrating the L radiation, of the element. To use the phraseology of optics, the X-ray spectrum of each element consists of two lines, one being of much shorter wave-length than the other. These lines are just as characteristic of the element as the lines in its optical spectrum. For each type of radiation the wave-length decreases as the atomic weight of the element increases. In fact the wave-length is inversely proportional to the square of the atomic weight.

More careful analysis has shown that each of the lines in the

X-ray spectrum of an element is really a " doublet," that is, it consists of a mixture of two kinds of radiation whose wave-lengths are nearly, but not exactly, the same. In other words, it consists of two lines quite close together, just as the D line of sodium is found to consist of two lines close together when a sufficiently powerful spectroscope is employed. The wave-lengths of some of these characteristic radiations are given in Table II.

It may be noted that the L radiation has not been detected in the case of elements of lower atomic weight than that of zinc. This is in part due to the fact that the wave-length would be so great that the rays would be absorbed in a very few millimetres of air, and would therefore be very difficult to detect. There is some evidence, however, for supposing that elements of lower atomic weight than forty-eight do not give rise to radia-tion of the L series. K radiation has not yet been obtained from elements of lower atomic weight than sodium, probably owing to the experimental difficulties of dealing with such very absorbable radiations. It is possible that other series of characteristic rays may be discovered after further investiga-tion. There is some evidence of their existence.

38. X-RAY SPECTRUM FROM AN ANTICATHODE.—If the radia-tion coming direct from the target, or anticathode, of an X-ray tube is examined it is found to give a continuous spectrum, that is to say it consists of a mixture of X-rays of every possible wave-length between certain limits. It may be compared to the light emitted from a metal ball which is raised to incan-descence in a Bunsen flame. If the tube is a " soft " one, requiring only a comparatively low potential to run it, and having, therefore, a small alternative spark gap, the radiation consists mainly of rays of comparatively long wave-length. As, however, the tube gets " harder " rays of shorter and shorter wave-length begin to make their appearance, and the spectrum extends further and further in the direction of decreasing wave-lengths. This is exactly analogous to the case of the heated solid. When the temperature is comparatively low only the red rays are visible in its spectrum ; as the temperature is raised the spectrum extends first into the yellow, and then in

succession into the green, the blue and finally into the violet, or in the case of a body such as the sun into the ultra-violet beyond.

A soft X-ray tube, therefore, resembles a body which is only just red hot, and the rays it gives out are confined to the longer wave-lengths. As the tube " hardens " shorter and shorter wave-lengths make their appearance, and the average wave-length of the radiation emitted decreases. In no case, however, is the radiation all of one wave-length, or homogeneous as we may call it. Its average quality changes, but it is always a mixture.

In addition to this " general " radiation the anticathode will also give out a certain amount of its own characteristic radiations superimposed upon the other, just as we might superimpose the sodium spectrum on the spectrum of a glowing solid by putting a little sodium salt in the flame. Under the conditions under which a tube is used in radiography the characteristic radiation is generally a small fraction of the whole.

39. ABSORPTION OF X-RAYS.—We have already given a general account of the absorption of the rays, but the matter is one of such importance in radiology that it is worth while to consider the matter more fully.

It is found that if we are dealing with homogeneous rays, that is to say, with rays having all the same wave-length, then the fraction of the incident radiation which is cut off by a screen of a given thickness of the same material is always exactly the same. Thus if an aluminium screen 5 mm. thick cuts off half the rays falling upon it, a second screen of the same thickness placed behind it will cut off half the rays passing through the first screen, that is to say, the two together will cut down the initial radiation to one-quarter. A third screen of the same thickness as the others would, in the same way, cut down the radiation transmitted through the two others to one-half, that is, the three screens would reduce the initial intensity of the rays to one-eighth of its original value. This is summed up mathematically by saying that the absorption of a homogeneous beam of X-rays is exponential.

It follows that theoretically no thickness of aluminium would cut off the rays completely, though the amount transmitted might be imperceptible. Thus, in the case we have been considering, seven of the screens (that is a total thickness of 3·5 cm.) would cut down the original radiation to $\frac{1}{128}$th of its initial value. This would probably be small enough to be negligible in practice. The absorption of a given substance for X-rays of a given wave-length can be expressed mathematically in the form—

$$I = I_o \, \epsilon^{-\frac{\lambda}{\rho}d},$$

where I_o is the intensity of the initial radiation falling on the screen, I its intensity after passing through a thickness d of the material, ρ is the density of the substance, and λ a constant (for rays of given wave-length), which is characteristic of the substance, and is known as its co-efficient of absorption. ϵ is a mathematical constant, the base of the natural system of logarithms. The values of λ for various wave-lengths are given for aluminium in Table II. Tables giving the values of ϵ^{-x} for different values of x are contained in collections of mathematical tables. They are known as exponential functions.

Suppose now we are dealing with a mixture of rays of different wave-lengths, as is always the case in practice. To make matters perfectly definite we will assume that our radiation consists of two definite wave-lengths, and that the one is cut down to one-half its intensity by an aluminium screen 0·5 cm. thick, while the other has its intensity reduced to one-quarter by the same screen. We will also assume that the intensities of the two kinds of radiations are initially the same. After passing through one screen the intensity of the first kind will be reduced to one-half and that of the second to one-quarter. The softer radiation in the transmitted beam is, therefore, only half as intense as the harder. After passing through a second screen of the same thickness the intensity of the harder rays is reduced to one-quarter and that of the softer to one-sixteenth. The latter is now only one-quarter as intense as the former. Thus the proportion of hard rays in the beam becomes greater and greater as the thickness of absorbing material increases. We can in this way filter out, so to speak, part of the softer radiation. It is obvious, however, from the example, that unless the thickness used is very large, and the consequent reduction in the intensity of the transmitted rays, as a whole, inconveniently great, or unless there is a great difference in the co-efficients of absorption of the rays, the " filtering " will be by no means complete. In general, however, we may say that, starting with a mixed radiation such as we get from the anti-cathode of an X-ray tube, the average hardness of the rays will be increased by passing them through screens of absorbing material, while any very soft radiation will be practically completely absorbed. This method is used in practice for eliminating the very soft rays which are largely responsible for the dangerous X-ray " burns."

40. Co-efficient of Absorption and Wave-length. —For substances of low atomic weight like aluminium, the co-efficient of absorption of the rays in the substance decreases continuously as the wave-length becomes smaller. The harder the tube and the shorter the consequent wave-length the greater the penetrating power of the rays. This applies to all substances in general, but every substance which is capable of emitting characteristic radiation shows also a kind of selective absorption for those wave-lengths which are capable of exciting its characteristic radiation. In other words, its co-efficient of absorption is exceptionally high for rays having wave-lengths somewhat shorter than that of its own characteristic radiation. It is obvious, from considerations of energy, that this must be so. The longer wave-lengths, as we have seen, do not excite the characteristic radiation of the element, while shorter wave-lengths excite it in considerable amounts. Now the energy of the secondary radiation must be derived from that of the primary beam. Hence, when the secondary is excited more energy must be withdrawn from the primary rays for the purpose. In other words, the absorption is increased. This is well shown by the figures given for copper absorbers in Table II. For very soft rays the thicknesses of copper and aluminium required to cut down the radiation to half value is

nearly the same. For rays of wave-length just greater than the characteristic radiation from copper, however, the absorption becomes very great, and only a small thickness of copper is required to cut down the radiation to one-half. From this point copper is a much better absorber than aluminium. Thus for the arsenic radiation an aluminium screen would require to be 7·8 times the thickness of a copper screen in order to produce the same absorption, for the molybdenum rays 8·4 times, and for the silver radiation 9·7 times the thickness. The absorptions in copper and aluminium no longer run parallel to each other.

This result can be used to estimate the quality (*i.e.*, the average wave-length) of the rays. Suppose a thin copper plate and an aluminium wedge were placed side by side and a beam of rays of the same wave-length as the characteristic arsenic radiation was passed through them. At the point on the wedge where the thickness of the aluminium was 7·8 times that of the copper the quantities of radiation transmitted through the two would be equal, and a fluorescent screen placed over them would appear equally bright on both halves. If, however, the radiation used had the wave-length of the silver radiation, a greater thickness of aluminium would be required to balance the absorption of the copper, and the balance would therefore be further away from the thin end of the wedge. The position on the wedge where the rays transmitted through the aluminium equal in intensity those which have passed through the copper plate is thus a measure of the quality of the radiation.

This is the principle of various penetrometers or qualimeters for estimating the penetrating power or hardness of the rays from an X-ray tube. Silver is generally used for the uniform plate in place of copper, as the absorption band of silver occurs with radiations slightly softer than that generally employed for radiography. The aluminium wedge is graduated according to some arbitrary scale. The position on the wedge where its transmission is equal to that of the plate is judged by allowing the rays to fall on a fluorescent screen.

As the selective absorption takes place for different wave-lengths in different elements it is possible by means of a suitable combination of screens to limit the radiation coming through them to a comparatively short part of the whole spectrum. It is possible that further research on these lines may enable us eventually to work with fairly homogeneous beams of rays.

TABLE II.

Element.	Wave-length of characteristic radiations (× 10^8 c.m.).				$\frac{\lambda}{\rho}$ (K series) in Aluminium.	Thickness required to reduce radiation to half.	
	K series.		L series.			Aluminum.	Copper.
	α	β	α	β		mm.	mm.
Al	8·36	7·91	—	—	—	—	—
Cr	2·30	2·09	—	—	136	0·05	0·05
Fe	1·95	1·77	—	—	88·5	0·08	0·07
Co	1·80	1·63	—	—	71·6	0·09	0·09
Ni	1·66	1·51	—	—	59·1	0·12	0·11
Cu	1·55	1·40	—	—	47·7	0·14	0·13
Zn	1·45	1·29	12·35	—	39·4	0·18	0·12
As	1·17	1·05	9·70	—	22·5	0·31	0·04
Mo	0·72	0·63	5·42	5·19	4·8	1·44	0·17
Ag	0·56	0·50	4·17	3·93	2·5	2·8	0·29
Sn	0·50	0·43	3·62	3·38	1·57	4·4	—
Sb	0·48	0·41	3·46	3·25	1·21	5·7	—
W	0·20	0·18	1·49	1·28	—	—	—
Pt	—	—	1·31	1·12	—	—	—

41. IONISATION PRODUCED BY X-RAYS.—We have considered this in an earlier section, and have seen how it may be used to estimate the intensity and quantity of the radiation. The results are only strictly comparable so long as we are using rays of the same mean wave-length. The ionisation produced in a gas is roughly proportional to the quantity of rays absorbed by it. Thus a penetrating radiation, being less absorbed in the gas, will produce relatively less ionisation than the same quantity of more absorbable radiation. The same defect attaches to all methods so far adopted for measuring the quantity of X-radiation, and further research in this direction is much to be desired.

An alternative method of measuring intensity or quantity is to find the time taken for the rays to discolour pastilles of barium platino-cyanide to a definite tint. If the rays are allowed to fall on one of these pastilles its colour gradually darkens, the extent of the action being proportional to the quantity of rays which have fallen on its surface. Thus the darkening of the pastille to a definite tint indicates the action of a definite quantity or dosage of the rays. The action, however, depends on the quality of the radiation, while the estimation of tints is a far less accurate operation than the reading of an electroscope or voltmeter. With proper appliances there is no reason why the ionisation method, which gives numerical results, should not be made more convenient, as well as more accurate, than the use of barium platino-cyanide pastilles.

CHAPTER VI

42. THE X-RAY TUBE.—The primitive Crookes' tube in which the rays were first discovered has many defects as a means of producing X-rays. The evolution of the modern X-ray tube has been the result of much thought and experiment, and the last word has not yet been said. The design is governed by many considerations, among the most important of which are (*a*) the obtaining of a point source of the rays ; (*b*) the provision of a suitable target for the cathode rays to fall upon ; (*c*) the removal of the large amount of heat generated in the tube by the discharge, most of which originates at the point struck by the cathode rays.

It is well known that a point source of light casts very sharp shadows, while an extended source gives very indistinct shadows. In the same way an extended source of X-rays gives very indistinct shadows. It is desirable, therefore, that the X-rays should come from a very small area. As the X-rays are given out at all points struck by the cathode rays the problem resolves itself into focussing the cathode rays on to a single point. This is effected by making the negative electrode or cathode, from which the cathode rays are emitted, concave. It is found that the cathode rays are always projected at right angles to the surface from which they come. Thus, if the cathode is part of a concave spherical surface the cathode rays, if they travelled in straight lines, would be brought to a focus at the centre of the sphere. As a matter of fact, owing to the action on the rays of the very intense electric field near the cathode, the focus is a little further from the cathode than the centre. If the target is placed at the point to which the cathode rays converge the X-rays will thus be emitted practically from a single point.

The impact of the cathode rays on a single point, however, is a considerable strain on the target. With the currents now used in radiography the energy in the cathode rays is very considerable, and even hard metals can be pitted by the force of the impact. The target is, therefore, generally fixed at a little distance beyond the exact focus, sufficient to prevent damage to the metal and yet, at the same time, not large enough to cause serious fuzziness in the shadows produced. Some makers, however, provide tubes with different degrees of sharpness in the focus, so that in cases where much fine detail is required a sharp focus tube can be employed, while in others, where the detail is not important, a tube with a broad focus can be used. The latter will, of course, stand a much heavier discharge without damage to the target.

In the Crookes' tube the cathode rays fell directly on the walls of the tube. The heat generated at the point of impact by a modern discharge would melt the glass almost instantaneously and thus ruin the tube. Glass is unsuitable also for other reasons. It has been found that the intensity of the X-rays generated by a given beam of cathode rays increases with the atomic weight of the target on which they fall. Thus, the target should be made of a metal of high atomic weight. Until recently platinum was generally employed. Tungsten, though a slightly less efficient radiator, has the advantages of being both harder and more infusible. It is, therefore, less easily damaged by the discharge, and it has very largely displaced platinum for use as the target (or anticathode, as it is called) in X-ray tubes.

It is in the methods of dealing with the heat produced that different makes of tube differ most. The first consideration is obviously to secure the rapid removal of the heat from the point where it is produced, and this is secured by making the actual target, whether of tungsten or platinum, part of a fairly massive block of metal. Copper is generally used as it is an excellent conductor of heat, and thus rapidly transmits the heat produced throughout its whole mass. If the copper has a fairly large surface it will radiate the heat it receives through the glass of the tube into the air. If the tube is not required for very heavy and continuous discharges the cooling produced by this radiation may be sufficient to keep it reasonably cool. A tube of this sort is, in fact, perfectly good enough for currents up to 3 milliamperes, and as considerably less work is involved in its construction it is much cheaper than tubes in which artificial cooling is introduced. One form of X-ray tube of this type is shown in section in Fig. 26.

The cathode C is of aluminium. and concave, for reasons stated above. It is supported on an aluminium rod, and is placed just at the neck of the large bulb B, in which the discharge takes place. The rod is connected with the external metal cap N by platinum wire fused through the glass. The anticathode is a block of copper bevelled at 45 degrees to the

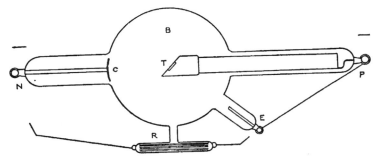

Fig. 26.—Diagram of an X-ray tube.

axis of the tube, and containing the target T let in flush with its surface, on to which the cathode rays are focussed. The paths of the cathode rays are indicated by the dotted lines. They will not, however, be visible unless the vacuum is lower than that generally used in actual work. The anticathode is prolonged by a copper tube which serves to support it, and also to radiate the heat. It is connected by a platinum wire with the external metal cap P.

X-rays are given out in all directions from T, but those going forward into the copper will of course be absorbed. Thus, only the hemisphere in front of the anticathode will be illuminated by the rays. This is shown graphically, when the tube is excited, by the vivid green fluorescence of the walls of the tube in front of the anticathode. The portions behind the anticathode remain dark.

Practically all tubes are fitted with an auxiliary electrode 三, known as the anode, which is connected by a wire outside the tube with the anticathode. The function of this is not altogether clear, but it seems to steady the discharge. R is a device for regulating the vacuum in the tube. We shall consider this later.

All the metal parts except the actual working surface of the cathode and anticathode are enclosed in glass mantles. It is found that when a discharge passes from a metal electrode minute particles of the metal are driven out by the discharge and settle on the glass walls of the discharge tube. This spluttering only takes place at the negative electrode, and is greater with metals of high atomic weight than of low. The effect of this coating of metal on the walls of a discharge tube is to render them partly conducting. It is then very difficult to make the discharge pass through the tube, and even if the discharge takes place it is very irregular. To prevent this cathodic spluttering the cathode is made of aluminium, for which the effect is very small, while all the metal parts which are not in actual use are protected by the glass mantles as already described. Spluttering will not take place from the anticathode so long as the current is passing through the tube in the right direction. If, however, the current reverses, making the anticathode negative, the spluttering may be considerable. In most forms of X-ray apparatus there is, as we shall see later, a certain amount of reverse current. It is, therefore, desirable to protect the anticathode with a glass mantle in the same way as the cathode.

The size of the bulb in which the discharge takes place is now generally from 6 to 8 inches diameter. With heavy discharges a large bulb works more steadily, and keeps its vacuum more constant than a small one, while the greater distance of the glass from the anticathode renders it less liable to become overheated. If the bulb is too large, however, it is apt to be unwieldy.

43. COOLING DEVICES.—The tube described may be taken as the standard form of X-ray tube, and the only important modifications are those connected with devices for cooling the

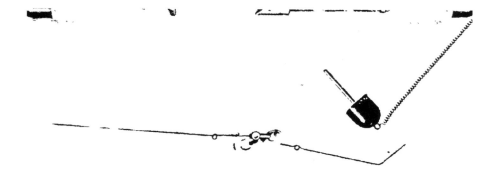

FIG. 27.—Air-cooled tube. (Mammoth, C. Andrews.)

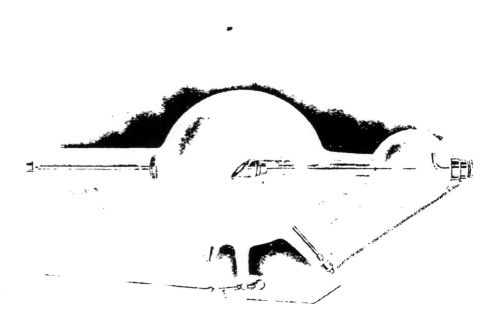

FIG. 28.—Water-cooled tube. (Rapidex, C. Andrews.)

anticathode. In some cases, of which the Mammoth (Fig. 27) is a good example, the copper tube supporting the anticathode is actually sealed into the walls of the discharge tube. The interior can then be placed in contact with the air outside without any fear of impairing the vacuum in the bulb. The circulation of air inside the tube materially assists the cooling. If necessary air can be blown through the space inside the anticathode by means of a mechanical blower.

In another type the interior of the anticathode is filled with water (Fig. 28). On account of its great thermal capacity water is a very efficient absorber of heat, and these water-cooled tubes can be run for long periods without overheating. In the case of very heavy discharges a constant stream of water can be kept circulating through the water chamber behind the anticathode. This is, however, not usually necessary. A certain amount of care is necessary with a water-cooled tube to make sure that it is never used without sufficient water in the reservoir, and also that the tube is clamped so that the water does not run out. The tubes are also more expensive than the simpler varieties. On the other hand, they are certainly capable of withstanding comparatively heavy currents for a considerable length of time without overheating and without altering their vacuum. Owing to the very rapid absorption of the heat by the water it is also possible to focus the cathode stream much more sharply upon the anticathode without fear of damaging the tube. It is thus possible to combine a point source of X-rays with a considerable current through the tube, and the radiograms produced by a well-made water-cooled tube are generally rich in fine detail. With reasonable care the life of the tube is very long, and many hundreds of radiograms can be obtained from a single tube. Thus, for use with installations capable of furnishing strong currents they are certainly to be preferred to the simpler types. If, however, the installation is a small one and not adapted for supplying currents of more than 3 milliamperes or so, they have no advantages over the simpler and cheaper types already described.

44. REGULATING THE TUBE.—It has already been stated

that when a discharge tube has been exhausted to the stage when the dark space begins to appear, still further exhaustion increases the potential necessary to maintain a discharge across the tube. The more complete the vacuum the greater the potential necessary to pass the current across the tube. This increase of potential increases the velocity of the cathode rays. If the potential required to run the tube is 100,000 volts, the velocity of the electrons making up the cathode rays is approximately $1·8 \times 10^{10}$ cm. per second. If the vacuum is increased so that a voltage of 200,000 volts is required to maintain the discharge, the velocity is increased to $2·2 \times 10^{10}$ cm. per second, or practically three-quarters of the velocity of light. Now the greater the speed of the cathode rays the shorter will be the wave-length of the X-rays which they give out when they strike the anticathode, and the greater will be the penetrating power of the rays. Thus the X-rays given out by a tube which requires, say, 120,000 volts to produce the discharge will penetrate through much greater thicknesses of material than the rays from a tube which can be run with only 60,000 volts. As the potential across an X-ray tube depends mainly on the degree of vacuum in the tube, it will be seen that the quality of the rays given out by an X-ray tube is determined by the degree to which the tube has been evacuated. A tube with a very high vacuum, and therefore requiring a large voltage to excite it, is termed hard, and the rays given off by it are called hard rays, and are very penetrating. On the other hand, a tube which contains gas at a slightly higher pressure, and which requires less voltage to excite it, is termed soft, and the rays given off are called soft rays. The terms " hard " and " soft " are relative only, and have changed their value considerably as high voltages and stronger tubes have come into use. For radio-graphic purposes we may consider a tube soft if the alternative spark gap is less than 5 inches. A tube with an alternative spark gap of more than 6 inches is considered hard. Tubes which have an alternative spark gap of less than 3 inches or more than 8 inches are generally unsuitable for radiographic work.

It is important to be able to regulate the vacuum in the tube in order to produce rays of the quality most suitable for the work in hand. Moreover, the pressure of the gas in an ordinary X-ray tube is far from constant. If the tube is over-run, that is to say, if too great a current is passed through it for too long a time, gas is given off from the electrodes which materially increases the pressure and makes the tube run soft. A new tube may be absolutely ruined in this way. On the other hand, if the tube is run with the normal current, it will be found that in process of time the tube becomes harder. This is possibly due to the molecules of gas, which become highly charged in the discharge, being actually driven into the walls of the tube. This reduces the pressure of the gas, and if no means of introducing a further supply of gas into the tube is provided the tube will eventually become too hard to use.

X-ray tubes are, therefore, provided with some means or other of increasing the pressure of the gas in the tube as required. The most usual way is to have a side tube R (see Fig. 26) sealed on to the main tube and containing asbestos soaked with some chemical which will liberate a small quantity of gas when heated. The most usual way of heating the asbestos is by an electric discharge. When the tube becomes too hard, a portion of the discharge is diverted through the asbestos until sufficient gas has been evolved to bring the tube back to its required condition. This kind of regulator is supplied in two forms. In the first form a single wire, making contact inside the tube with the asbestos, is provided which can be turned so as to approach the cathode terminal of the tube. When the end of the wire is sufficiently near the terminal a portion of the current passes along this wire, through the asbestos and through the tube to the anode. When the softening has been sufficient the wire is turned back to a sufficient distance from the terminal to prevent any further sparks passing between them.

A better type, working on the same principle, is the one shown in Fig. 26. This is fitted with two wires, the one making contact with the cathode as before and the other with the anode of the tube. The latter is usually adjusted so as to leave a gap of about a quarter of an inch between the end of the wire and the cap of the anode. The other wire is thrown well back until the tube needs softening. When it is required to soften the tube this wire is brought near the cathode cap, until spark-ing occurs. The current then passes through the asbestos and the two wires, liberating the requisite amount of gas. The advantage of this arrangement is that it provides a complete circuit for the current. If the tube has been allowed to become very hard it is sometimes impossible to get the current to pass through the first type of softener, at any rate without applying a potential which may spark through and ruin the tube.

The supply of gas condensed on the asbestos is not inexhaus-tible, but it is generally sufficient to outlast the tube. In some cases, however, where the tube has been used with special care, and unusual good luck, the gas supply gives out before the tube breaks. The convenience of the method, which allows of the tube being softened even while it is running (the wires can be manipulated with a long boxwood rule or other non-conducting rod), in practice far outweighs this disadvantage, and this is the form of softener usually provided by the makers of X-ray tubes, unless some other form is specified.

45. Osmosis Regulations. — Another form of regulator sometimes used consists of a platinum or palladium tube sealed at one end but

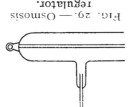

Fig. 29. — Osmosis regulator.

opening into the X-ray tube at the other (Fig. 29). The metal tube, being sealed, maintains the vacuum perfectly when cold. When heated, however, these metals have the property of allowing hydrogen to diffuse through them. Thus if the end of the tube is heated by a flame, the hydrogen in the flame diffuses slowly through the red-hot metal, and thus increases the pressure of the gas in the tube. On removing the flame the diffusion at once ceases. As the gas is introduced from outside the tube the regulator will continue to work in-definitely. The process is, however, less convenient than the one with the asbestos tube, and perhaps requires rather more practice to produce exactly the degree of regulation required. Special devices have been invented to allow of the flame being operated from a distance and while the tube is working, but, in general, it is desirable to avoid all unnecessary complications in X-ray appliances. The osmosis regulator is, however, very useful in the case of tubes, such as those used for therapeutic purposes, which require frequent softening, as the supply of gas is inexhaustible.

Other forms of regulators have been devised, such as the Bauer air valve. This consists of a delicate valve sealed by a thread of mercury and communicating with the outer air. By depressing the column of mercury the valve opens and allows a small quantity of air to enter the tube. The slight increase of pressure in the tube forces the mercury column back and again seals the tube. This type of regulator is unsuitable for tubes having tungsten anticathodes, which materially limits its usefulness. It does not appear to have any marked advantage over the much simpler and less costly osmosis regulator.

46. '' HARDENING '' TUBES.—So far we have been considering only the methods of softening a tube which has become too hard for our purposes. The reverse process, that of hardening a tube which has become too soft, is much more difficult, and in the case of a tube which has become very soft is often impossible. In this case the only thing to do is to send the tube back to the maker for re-exhaustion.

In general, the necessity for hardening a tube arises only from accidental or careless overrunning of the tube. We have seen that an X-ray tube, when run at its proper current, eventually becomes harder, owing to the absorption of the residual gas by the walls of the tube. If, however, too much current is passed through the tube, especially in the case of a new tube, the electrodes become unduly heated, and will then liberate their occluded gas. A newly made tube, in which the electrodes still contain their normal amount of absorbed gases, may very easily be ruined by too great a current. It is obvious that, since the walls of the tube absorb gas, while the electrodes, when heated, give out gas, there will, for a new tube, be some current for which the gas given out by the electrodes is equal to that absorbed by the glass in the same time. In this case the pressure in the tube, and thus the hardness of the rays, remains practically constant. This critical current varies with the mode of exhaustion of the tube, but is generally about 3 or

4 milliamperes. It is obviously most convenient and most economical to use the tube, as far as possible, with the critical current.

If, however, owing to overload, the tube has become slightly too soft, it can be hardened by running it for some time with a current appreciably smaller than the critical current. The tube should be allowed to stand until perfectly cool (if it has been overheated), and a current of from 1 to 1½ milliamperes passed through it for ten minutes. It should then be allowed to rest for an hour or so, and again tested. If still too soft, the process may be repeated. The method can, of course, also be used in the case of a tube which has become over-soft by careless use of the regulator.

If the softening has been at all great the process is apt to be very tedious, and may be unsuccessful. It is always worth trying, however, before sending the tube back to the maker. It is, in any case, the only method generally available. If, however, the tube contains hydrogen as its residual gas, that is to say, if the tube was filled with hydrogen before being exhausted, and is fitted with an osmosis regulator, the osmosis regulator can be used to take gas out of the tube, as well as to put it in. In the latter process the palladium or platinum tube is heated by a flame. As the flame contains hydrogen at a greater pressure than that in the tube the hydrogen diffuses through the metal into the tube. If, however, we heat the tube, not with a flame, but by means of a spiral of wire maintained at a red heat by passing a current through it, there will be no hydrogen outside the metal tube, and the hydrogen inside the X-ray tube will diffuse out, since its pressure inside the tube, though very small, is greater than the pressure of hydrogen outside, which is zero. A device of this kind is fitted to the MacAlister-Wiggin hydrogen tube. In any tube which is regulated by means of osmosis the residual gas will, after the tube has been in use for some time, consist mainly of hydrogen, as hydrogen is the gas introduced each time the tube is regulated, and the same method may be applied with some chance of success. A spiral of fine platinum wire is slipped over the end of the osmosis tube, so as not quite to touch it at any point. The spiral is raised to a white heat by the current from two or three accumulators. The effect on the tube may be tested from time to time, but owing to the very small pressure of hydrogen the process is usually very slow. It is obvious, therefore, that great care should always be taken to avoid over-softening the tube either by overrunning or by a careless use of the softener.

47. THE COOLIDGE TUBE.—A new form of tube has been devised by Coolidge which entirely overcomes the difficulties due to alterations in the pressure of the residual gas. In this tube the gas is completely removed from the tube and from the metal electrodes by a prolonged process of heating and exhausting. In fact, the tube is so completely evacuated that it is quite impossible to pass a discharge across it in the usual way. In the ordinary type of tube the current is conveyed from one electrode to the other by the action of the molecules of the residual gas. In the Coolidge tube the number of these is so small that no appreciable current can be transmitted by their means. To overcome this difficulty the cathode is made of a closely wound spiral of tungsten wire which can be raised to a white heat by means of an electric current from an insulated battery of storage cells. A metal when heated is found to emit negative electrons, or thermions as they are often called, which are identical with those con-stituting the cathode stream in an ordinary discharge tube. The quantity of these thermions emitted by the spiral in-creases rapidly with the temperature, and can be increased or diminished at will by raising or lowering the temperature of the spiral. If now the tube is excited in the usual way while the spiral is glowing, the thermions set free from the latter will be projected by the electric field across the tube, the velocity which they acquire being proportional to the square root of the potential difference between the electrodes. It is found that the thermions can be focussed into a narrow beam by surrounding the spiral with a metal tube. The beam can then be made to impinge on an anticathode in the usual way.

It will be seen that the tube is under very much better control than tubes of the ordinary variety. The hardness of the rays given out is entirely fixed by the potential difference which we chose to apply between its terminals. Increasing the potential given by our coil or transformer automatically increases the hardness of the rays, while decreasing the potential automatically "softens" the tube. All difficulties connected with regulators and so on are, therefore, removed. At the same time the quantity of the X-rays given off, which depends on the number of thermions in the cathode stream, can be adjusted independently of the potential by increasing or lowering the temperature of the spiral.

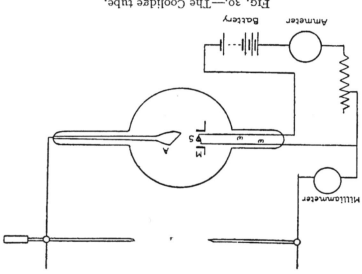

Mill_Amm_eter

Ammeter   Battery

FIG. 30.—The Coolidge tube.

The present form of Coolidge tube is shown diagrammatically in Fig. 30. S is the spiral of tungsten wire, surrounded by a molybdenum tube M. The anticathode A, which also serves as anode, is a massive block of tungsten supported by a molybdenum rod which is welded to it. The current for heating the spiral is conducted through the two wires *ww*, fused through the end of the tube. One of these wires is joined to the negative terminal of an induction coil, the other terminal of which is connected directly on to the anode A. It will be seen that the whole of the battery providing the heating current is raised to the same potential as the cathode of the tube, and must, therefore, be carefully insulated. It is as dangerous to touch the battery or any of its connections while the tube is working as to touch the terminals of the induction coil itself. An ammeter should be included in the heating circuit to measure the heating current. The current through the tube is measured by a milliammeter in the usual way.

No cooling device is required. It is impossible to injure the massive anticathode, and as the latter has been completely freed from gas, overheating will not cause any alteration in the vacuum, and it is stated by the makers that the tube may safely be run with the anticathode at a white heat. Care must of course be taken that the glass does not become hot enough to soften, or the tube will puncture. It may be noticed that so long as the anticathode is not at a temperature comparable with that of the spiral, current can only pass through the tube when the spiral is the cathode. If the spiral becomes positive, the electric field will simply drive the thermions back into the spiral, and no current will cross the tube. The Coolidge tube acts as its own rectifier.

These numerous advantages would seem to point to its general adoption in the near future, at any rate, in large installations. Its main disadvantage (besides the cost, which is very high) is the necessity for the auxiliary battery and apparatus. The accumulator cells need careful attention and frequent charging, and of course must be carefully insulated. The difficulties of regulating a tube of the ordinary type, though somewhat formidable at first, rapidly become less with practice, and for smaller installations it is possible that the extra apparatus needed by the Coolidge tube may be thought to outweigh its many undoubted advantages. At the moment, while it has been adopted by many of the leading radiographers, its use is by no means general, and many radiographers have stated that the radiograms obtained with the Coolidge tube are lacking in the fine detail shown by those obtained with an ordinary tube. This is due to the fact that X-radiation is emitted not only from the focus, but to a quite appreciable extent by the whole of the tungsten anticathode, and even by part of the molybdenum support.

48. High Tension Rectifiers.—Owing to its construction

it is absolutely necessary that the current shall pass through an X-ray tube in the proper direction, that is, from the anticathode to the aluminium cathode.   If the current passes in the reverse direction, not only is it useless for radiographic purposes, but the tube itself becomes damaged rapidly by the metallic sputterings from the anticathode which now functions as the cathode.   It is also possible that the stream of cathode rays, which will now be given off from the anticathode and fall upon the walls of the tube, may be sufficiently intense to puncture the glass.

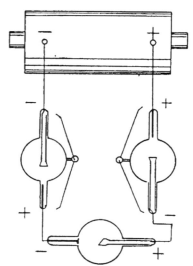

As we have seen (§ 24) the discharge produced by an induction coil is not entirely in one direction. In addition to the main current which we use for exciting our tube the coil also produces more or less inverse current, that is, current in the opposite direction to the principal discharge.   This inverse current is generally much weaker than the direct current.   It is, however, often sufficiently large to cause trouble and must be suppressed.

The usual method is to insert a rectifier or valve in series with the X-ray tube.   These valves have the

FIG. 31.—Method of connecting high tension rectifiers.

property of only transmitting current in one direction.   It has been found that if a vacuum tube is furnished with two electrodes of very unequal area and exhausted to a suitable low pressure the discharge will pass through the tube very easily (that is, at a comparatively low voltage) when the large electrode is made the cathode or negative terminal, but only with great difficulty if the small electrode is the cathode.   The effect is increased if the small electrode is placed in a confined part of the vessel, say, for example, in a narrow side tube.   This principle has been applied in various ways.   In the usual Villard valve the cathode consists of a large spiral or tube of aluminium, while the anode is a small metal plate or rod in a narrower part of the tube.   In using the valve the spiral must be connected directly to the anode of the X-ray tube, or if preferred, the small metal plate may be connected directly to the cathode of the X-ray tube. It is quite usual, in fact, to have two valves in the circuit, one on each side of the X-ray tube, as shown in the diagram of Fig. 31.

The pressure of the residual gas in the valve tube should be much greater than that employed in X-ray tubes, so that a mparatively small difference of potential will be sufficient to in the valve. Otherwise very considerable resistance will be added to the circuit and the output of rays will be reduced. The valve tube should have an alternative spark gap of about cm. At this pressure the whole of the tube when working is seen to be filled with a bluish glow, except for the Crookes' dark space which should be visible round the cathode. As valve tubes become hard in the same way as X-ray tubes and for the same reasons, it is necessary to have them fitted with some form of regulator, as described in § 36. An osmosis regulator is greatly to be preferred, as there is no limit to the amount of gas which can be introduced by it, but the asbestos type is the more common.

Other forms of valve differing only in the position and shape of the electrodes have been constructed. In each case the larger of the two electrodes is the cathode. In the Lodge valve the pressure of the residual gas is maintained constant by introducing red phosphorus into the valve tube before sealing off. The red phosphorus has a small but appreciable vapour pressure which is of the right order for the action of the valve. The pressure is maintained at this value by the gradual evaporation of the phosphorus.

It is often possible with a good coil to work without a rectifier, at any rate for currents of 1 or 2 milliamperes, though different tubes seem to differ very much in this respect. A well-seasoned tube will often run quite satisfactorily without a rectifier, while under the same conditions a new tube may show so much inverse current as to be quite useless. In any case, valve tubes are so cheap that it is unwise to have an installation without one. In case of emergency a spark gap consisting of a fine point facing the centre of a large, flat metal disc, and

placed in series with the X-ray tube, will often produce a considerable degree of rectification, at any rate for small currents.
The spark gap functions very much like the valve tube, the
current passing much more readily across the gap when the
point is positive than when it is negative.   To obtain the best
results the width of the spark gap should be adjustable.   The
best position can then be found by trial.   A gap of from 1 to
2 cm. is generally required for a current of 1 milliampere.
Specially constructed spark gaps have been placed on the
market for this purpose, but they are somewhat noisy, not very
efficient in practice, and in any case will only rectify comparatively small currents.

Modern installations are now often fitted with mechanical
means of rectifying the current furnished by the coil.   These
will be described later (§ 58).

49. THE OSCILLOSCOPE.—The presence of inverse current in

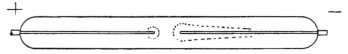

FIG. 32.—The oscilloscope.

the circuit can generally be detected by observing the fluorescence on the walls of the X-ray tube.   When all is going well
the hemisphere in front of the anticathode glows with a steady,
greenish light, while the inside of the tube is dark except for a
little bluish light around the anode.   If, however, there is
inverse current, the glass behind the anticathode begins to
show flickers of greenish fluorescence, while patches of the
blue glow appear in various parts of the tube.   With a little
experience it is possible not only to detect the presence of inverse
current, but also to estimate approximately the hardness of
the tube, by observing its appearance while working.

In many cases, however, the tube is enclosed in a box, so that
it is impossible to watch it at work.   The presence of inverse
current can then be detected by including an oscilloscope in
series with the X-ray tube.   The oscilloscope (Fig. 32) consists
of a small glass tube furnished with two long aluminium wires,
which lie axially along the tube and are separated at their ends

by a gap of 2 or 3 mm. The tube is partially exhausted of air, and sealed. When a current is passed through the tube the negative wire is covered with a bluish glow, while the positive wire is dark except for a bright spot of light at its extreme end. The direction of the current can thus be detected. If inverse current is present in the circuit each wire in turn acts as the cathode, so that the glow appears on both wires. The extent of the glow along the positive wire is a rough measure of the amount of inverse current in the circuit. These tubes are very useful to check the working of the circuit, and as they are very cheap and give no trouble, one should always be included in circuit with the X-ray tube.

CHAPTER VII

THE PRODUCTION OF HIGH TENSION CURRENTS

50. THE PRODUCTION OF HIGH POTENTIAL CURRENTS.—As we have seen (§ 44), to excite X-rays of the penetrating power required for radiography we require to be able to apply to the terminals of the X-ray tube a potential difference of at least 100,000 volts. This is far greater than the potential at which electric current is generally applied for other purposes. The electromotive force of an accumulator cell is about 2·1 volts. The current supplied for lighting purposes has generally a pressure from 100 to 300 volts, though it is frequently generated at pressures of from 2000 to 10,000 volts. Even these potentials are, however, far too small for our purpose, and we need some means of increasing very largely the voltage of our supply before we can use it for radiography.

High tension currents may be generated—

(a) By influence machines, of the type of the Wimshurst.

(b) By the induction coil.

(c) By a high tension transformer.

The action of the influence machine has been already explained (§ 8), and modern forms seem to have been used with success for radiographic work in France. The current supplied is unidirectional (there is no inverse current) and increases with the rate at which the machine is turned. While at present it does not seem capable of furnishing the very large discharges which are used in modern radiography, and insulation difficulties are apt to be serious in a climate like that of Great Britain, none of these difficulties should be insuperable, and if they can be overcome, the many obvious advantages of the machine would rapidly bring it into general use. The only two methods in common use at present are, however, the induction coil and the transformer.

51. The Induction Coil.—The induction coil, as we have seen (§ 24), consists of two separate circuits, a primary circuit and a secondary circuit. The primary circuit is wound in a spiral upon a long straight core of soft iron, and, as it must be capable of carrying a strong current without becoming overheated, it is made of comparatively thick copper wire. As no very large difference of potential exists between the separate turns of the primary winding, it is sufficient to have them insulated from each other by the usual silk or cotton covering employed for insulating wire.

The secondary circuit consists of a very large number of turns of fine wire, wound upon an ebonite tube, into which

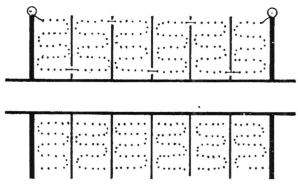

Fig. 33.—Sectional winding of an induction coil.

the primary winding, with its iron core, exactly fits. The secondary winding thus surrounds the primary so that the two are as close together as possible. As the potential differences in the secondary circuit will be very large it is necessary that the insulation should be as strong as possible. This is effected by coating the wire with insulating wax as it is wound on the coil, and, further, by embedding the whole of the secondary in wax after winding. Even so, however, there is grave risk of the insulation breaking down if, as in some of the cheaper coils, the secondary is simply wound from one end of the ebonite tube to the other and back again. In this case the full difference of potential of the coil will exist between the innermost and the outermost layer, that is to say, across the thickness of the coil, and a breakdown is almost inevitable.

To overcome this difficulty, high class coils are invariably wound in a number of flat sections, each section being insulated from the next by a layer of insulating material of sufficient thickness to stand the tension between the faces of the adjacent sections. The method of winding is sufficiently indicated by Fig. 33. It will be seen that the different sections may be regarded as a series of secondary coils upon the same axis, their ends being joined together so that the potential increases steadily from one end of the system to the other. In this way the potential difference between successive layers of winding is very much reduced, and the full potential is only developed at the extreme ends of the coil.

The secondary is generally covered with a thick layer of wax, which is itself protected by a covering of ebonite. In the majority of coils this is simply a thin sheet of ebonite laced round the coil. In the coils used in " wireless " telegraphy the secondary is usually enclosed in an ebonite tube. This is certainly a preferable arrangement, as it adds not only to the insulation but also to the mechanical strength of the structure.

The potential difference between the ends of the secondary increases with the number of turns of wire in it. In order to have a large number of turns without making the coil too heavy and bulky the wire used in the secondary is comparatively thin. On the other hand, if the wire is very thin its resistance will be very great, and thus, though the potential may be very high, the current through the tube will be small. For this and for many other reasons, which it is impossible to go into here, the construction of every coil is a compromise between various factors, and no coil can be equally suitable for all classes of work and for all conditions. Some adjustment can be made by winding several distinct primary coils on the same core, and this is now generally done. The complete theory of the induction coil has not yet been arrived at, but the different makers have accumulated a considerable amount of practical experience, and will generally be able to suggest the best coil for a given set of conditions.

It must be remembered that, while a potential of at least 50,000 volts is required to pass any current through an X-ray tube of the hardness necessary for radiography, and that voltages up to 120,000 volts may be required for a very hard tube, the factor governing the output of the rays is the amount of current which the coil is capable of sending through the tube. This depends not only on the induced electromotive force but also on the resistance of the secondary circuit. Some of the older coils wound with an enormous number of turns of very fine wire produced very high potentials. The celebrated giant induction coil wound by Spottiswood, which contained 280 miles of wire, gave a spark of 42 inches in air. Such large spark gaps are not needed in radiography, and the resistance of such coils makes the resultant secondary current very feeble. A coil giving a spark of 16 inches is quite sufficient for all purposes if wound so that it is capable of giving the necessary current. This depends, in the first place, on the power which can be applied with safety to the primary coil without overheating it ; in the second place, on the resistance of the secondary coil ; and, in the third place, on the efficiency of the coil. It is advisable, when buying a coil, to have some guarantee not only as the maximum spark which the coil will give but also as to the current which it will pass through, say, an 8-inch spark gap in air or through a tube in a standard condition. A good 16-inch coil is capable of producing more current than can safely be passed through any X-ray tube except for a very short fraction of a second.

52. THE INTERRUPTOR.—It is obvious, from what has been said in § 27, that in order to have anything approaching a continuous current through our X-ray tube we shall require some means of making and breaking the primary circuit a very considerable number of times in every second. Any device for producing this continual make and break is known as an interruptor. The almost innumerable forms of interruptor on the market can be reduced to three main types.

53. " DIPPER " BREAKS.—This is perhaps the simplest type of break. In principle it consists of a piece of metal which can be made to dip into a pool of mercury. If the primary circuit is cut and one end connected to the metal and the other to the mercury it is evident that the current will flow when

the metal dips into the mercury, and stop when it is with-drawn.

A simple "break" on these lines is figured in Fig. 34. It consists of an axle A having a number of projecting spokes. The axle is rotated by a simple form of electromotor which may be worked by the primary current itself. As the axle rotates each spoke in turn makes contact with the surface of a pool of mercury M. The primary circuit is made each time a spoke enters the mercury, and is broken when it leaves. The number of breaks per second is determined by the rate of revolution of the axle, while the duration of the contact can be adjusted by raising or lowering the level of the mercury so as to give the best results. This break, though very simple, gives good results when the primary current used is not too great.

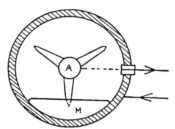

FIG. 34.—" Dipper " break.

In all interruptors in which mercury is used, the spark which always passes when the circuit is broken tends to cause chemical combination between the mercury and the oxygen in the atmosphere. Thus, if the break is filled with ordinary air it rapidly becomes clogged by the formation of solid mercury oxide. This difficulty is overcome in practice by replacing the air either with some gas such as hydrogen or ordinary coal gas, which has no action on the mercury, or by covering the mercury with a layer of alcohol or paraffin oil to keep the mercury from contact with the air. The first method is much to be preferred. It entails a little more trouble in the manufacture of the interruptor, as all the joints have to be made gas-tight. It is, however, clean and convenient in use. If a gas supply is laid on, the break is connected direct to a gas tap and the air swept out by allowing the gas to flow through the break for a few minutes. The exit tap is then closed, the break being left connected with the supply in order to keep up a slight pressure in the reservoir, and thus prevent air from leaking in. If there is no gas supply, gas can be bought in cylinders and a gas bag used to regulate the pressure. One cylinder of gas will last for months. With gas in the break e mercury keeps clean for a considerable period, and can easily be freed from oxide when necessary by passing it through a pin-hole in a paper funnel.

The breaks having liquid dielectrics become dirty much quicker than those with gas, while the process of cleaning the break and separating the mercury from the emulsified liquid in the break is one which, if once tried, will not soon be forgotten.

54. JET BREAKS.—The dipper breaks have, in spite of their simplicity, been largely replaced by breaks in which the current is made and interrupted by means of a jet of mercury. This type gives a cleaner and quicker break of the current than the dipper, the blades of which are apt to drag a little mercury along with them on leaving the liquid, and thus prolong the contact. They are also capable of interrupting much larger currents. Every maker of apparatus catalogues one or more forms of jet break. The principle is indicated in

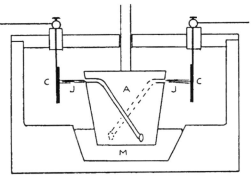

FIG. 35.—Principle of the mercury jet break.

Fig. 35. The break has a central cone A, which is rotated at a fairly high speed by a small electromotor. The lower end, which dips in a pool of mercury M, has a small hole facing in the direction in which the cone rotates. When the cone is made to rotate rapidly the inertia of the mercury forces it into the hole and up the tube so that it shoots out at the upper end in the form of a horizontal jet. The rotating cone thus acts as a kind of very simple centrifugal pump. A copper plate is placed near the cone so that the jet strikes it in the course of its revolutions, and one end of the circuit is connected to the copper plate and the other to the reservoir of mercury. Since the jet of mercury is a conductor contact is made every time the jet falls on the plate and broken as soon as it leaves it. The break is very sudden and the apparatus works very well in practice.

In modern types, illustrated in Fig. 35, the cone is bored with two tunnels so that two jets JJ emerge at opposite ends of a diameter. Two copper plates CC are employed and adjusted so that the jets strike them simultaneously. The two plates are connected to the opposite terminals of the circuit and contact is made between the plates by the two jets. The circuit is obviously made and broken twice in each revolution of the cone. The break must, of course, be filled either with coal gas or some liquid dielectric to avoid oxidation of the mercury. The duration of each contact is governed by the width of the copper plates ; a narrow plate implying a short contact, a wide one a prolonged contact. In some forms the copper plates are arranged so that they can be removed and replaced by others of different width. In this way the time of each contact can be regulated to suit the class of work to be done and the current available.

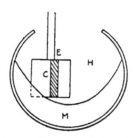

This form of break works very well and gives little trouble. It needs taking down occasionally in order to clean the mercury and clear the jets of any particles of dirt which may have become lodged in them.

FIG. 36.—Principle of the centrifugal break.

55. CENTRIFUGAL INTERRUPTORS.—A mercury interruptor of somewhat different type is also used. The mercury is contained in a small spherical iron chamber which is rotated at high speed by an electromotor. The mercury is set in rotation by the container, and owing to the centrifugal force forms a layer up the sides of the bowl. The rotating mercury forms one electrode, while the other is a copper strip, or series of copper strips, carried on a vertical cylinder which rotates in the same direction as the mercury. The cylinder is placed excentrically in the bowl, as shown in Fig. 36. The distances are adjusted so that when the copper strip is nearest to the bowl it is immersed in the moving layer of mercury. As it rotates it is carried out of the mercury, thus breaking contact. As the mercury and the copper are both moving in the same direction the break is particularly sharp. This form of break

has the additional advantage that by moving the cylinder nearer to or further from the bottom of the bowl the duration of each contact can be adjusted while the break is actually running.

So far it does not seem to have been found possible to adapt this type of break to work in gas. Paraffin is used to prevent oxidation, and the mixture emulsifies somewhat, in spite of the centrifugal action, so that the break requires an occasional cleaning out. On the other hand, it will interrupt rather larger currents than breaks of the jet type.

56. ELECTROLYTIC INTERRUPTORS. — Although mercury breaks work very steadily and quietly and are most suitable for general work, they are not capable of carrying the very

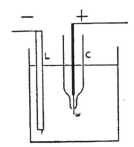

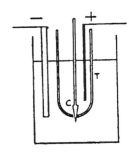

FIG. 37.—Wehnelt break.   FIG. 38.—Simon break.

large primary currents which must be used if very intense rays are required. For intense discharges a different type of break, working on quite different principles, is employed. This is the electrolytic break, of which there are two main types —the Wehnelt, and the Simon or Caldwell interruptor. The principle of the two kinds is much the same, but the method of application is different.

Wehnelt Interruptor.—The Wehnelt interruptor consists of a jar containing sulphuric acid of a specific gravity 1·2. In this are placed two electrodes. The negative electrode L consists of a large plate of lead ; the positive electrode of a piece of platinum wire w, about 1 mm. or 2 mm. in diameter. The platinum wire is enclosed in a porcelain cylinder C, which is usually tapered towards the bottom, where it is closed except for a small hole, slightly larger than the wire through which the latter projects. The wire is carried on a screw so that the length of it projecting through the hole can be easily adjusted. The break is shown in section in Fig. 37.

When the current is passed, the acid in the immediate neighbourhood of the platinum point becomes strongly heated, owing to its high resistance. In fact, it becomes so hot that it vaporises, forming a bubble of steam round the point. The steam being non-conducting breaks the current. As soon as the current is broken, however, the steam is condensed by the large mass of cold liquid around it and the bubble collapses, allowing the liquid to make contact with the wire again. Thus the current is automatically made and broken, and, if the break is properly adjusted, the interruptions are very rapid. As many as 2000 interruptions per second can be obtained if the amount of wire exposed is very small.

For heavy discharges the Wehnelt break is very convenient. As can be seen, it is very simple, and there is little to go wrong. A single Wehnelt interruptor is capable of interrupting efficiently very considerable currents, while for still larger currents three of these breaks can be connected in parallel. For continuous work it may be necessary to arrange to keep the acid cool by running a stream of water through lead pipes immersed in it. The acid should be tested from time to time to see that its specific gravity has not altered. Any change in the strength of the acid materially reduces the efficiency of the break.

The Wehnelt requires a source of supply of a fairly high voltage, preferably between 60 and 80 volts, and it does not work well with currents of less than 10 amperes. Owing to the great intensity of the discharges which may be produced by its means, care is required in its use, as otherwise the tube may easily be overheated, or in extreme cases the anticathode may be pitted or pierced.

The Simon Interruptor.—This interruptor, which is sometimes known as the Caldwell interruptor, differs from the Wehnelt in using a small lead plate in the anode, and thus eliminating the expensive platinum wire. The small lead anode is enclosed in a porcelain tube T (Fig. 38), which is pierced at the bottom by a circular hole.   A porcelain cone C carried on a long porcelain rod fits into the hole, the size of which can be adjusted by raising or lowering the cone.   The density of the current rises to a very high value when the hole is small, as it all has to pass through the small quantity of acid in the hole.   The heat developed at this point evaporates the liquid and thus breaks the current. The frequency of the interruptions can be adjusted by varying the size of the hole by means of the porcelain cone.   The specific gravity of the acid should be 1·2, as in the case of the Wehnelt break, and a voltage of at least 100 volts is required to work the break.   Given a sufficiently high voltage the break works very steadily and well and gives little trouble.

Both breaks suffer from the disadvantage that they are exceedingly noisy, the continuous formation and collapse of the bubbles making a rattling sound which is somewhat disconcerting.   When the break gets at all hot acid fumes are also evolved which may be unpleasant.   These troubles can be minimised by enclosing the break in a felt-lined box.   It is better, however, if possible, to arrange for the break to be installed in a separate room.

57. ALTERNATING CURRENT INTERRUPTORS.—All the interiptors described require direct current supply.   The Simon break will work on an alternating circuit as it acts as its own rectifier, but its use in this way cannot be recommended for X-ray work.   Numerous interruptors have been designed to work on the alternating circuit, but in spite of a large expenditure of ingenuity none of them attain the efficiency of the direct current interruptors.   If the only supply is alternating current (as often happens), it is much better to install a small motor-generator consisting of an alternating current motor arranged to drive a continuous-current dynamo geared on the same shaft. This combination has a very high efficiency, and is very simple to work, requiring no attention except an occasional oiling.

The alternating current interruptors are mostly of the rotating mercury jet type.   The cone is rotated by what is known as a synchronous motor, that is to say, a motor the revolutions of which correspond with the alternations in the electric current supply.   It is thus possible to arrange that the current shall always be flowing in the same direction in the circuit during the time that the mercury jet is in contact with the plate. The motor has to be started by a turn of the hand, and when it has picked up the proper speed (which can be ascertained by the uniform hum it emits), the current through the break can be switched on in the usual way.

58. MECHANICAL HIGH TENSION RECTIFIERS.—We have already considered the inverse current produced by an induction coil when the contact is made, and some of the devices employed for eliminating it. It is possible, however, to adapt a mechanical arrangement to a mercury break which will automatically effect the same object.

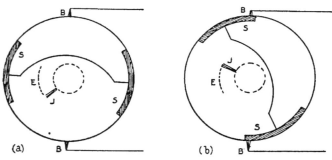

FIG. 39.—Mechanical high tension rectifier.

A large ebonite circular disc is fastened on to the axle of the motor driving the break so that the disc revolves with the mercury jet. Two copper strips are let into the disc at opposite ends of a diameter and connected by a wire. The secondary circuit is broken and the two ends brought to two brushes which make contact with the disc at opposite ends of a diameter. Thus when the disc revolves so as to bring the brushes into contact with the copper strips the secondary circuit is complete; when the disc moves onwards so that the brushes rest against the non-conducting ebonite the secondary circuit is broken.

Suppose now that the strips are arranged as in Fig. 39. The full lines in the figure represent the disc with the metal strips SS and the secondary brushes BB, while the dotted lines represent the copper electrode E of the break and the mercury jet J, which are, of course, vertically below the disc. At the moment represented by Fig. 39, *a*, the jet is about to make contact with the electrode E, and thus an inverse electromotive force will be set up in the coil. But, as will be seen, the secondary circuit at this time is broken and thus no current can flow through the tube.

As the disc and the jet rotate the moment comes, as represented in Fig. 39, *b*, when the jet is about to leave the electrode and thus break contact. At the same time the copper strips have revolved so as to touch the brushes. Thus, when the direct current is induced in the coil the secondary circuit is complete and the current flows without interruption through the tube.

This type of mechanical rectifier is due to Dr. Morton. A more complicated design, due to Mackenzie-Davidson, actually reverses the connections of the X-ray tube to the coil at the moment when the current is made, so that the inverse current is directed through the tube in the proper direction, thus not merely suppressing but actually utilising the inverse current. In good coils, however, the inverse current is so small that it is questionable whether the extra discharge thus gained is worth the extra complication in the mechanical arrangements.

In practice these mechanical rectifiers work very well, producing a discharge which is quite free from any inverse current. It is thus unnecessary to have a valve tube in the circuit if one of these mechanical rectifiers is employed.

59. High Tension Transformers. - As we mentioned in a previous chapter, a step-up transformer can be used to furnish very high potentials if the ratio of the number of turns of wire on the secondary to the number on the primary is sufficiently large. If the transformer is supplied with alternating current, then an alternating current of high electromotive force will be produced in the secondary. No interruptor is required, since the current is itself alternating.

In order to use the high tension current of a transformer for X-ray production it is necessary to rectify it. This can be done either by breaking the secondary circuit when the current is in the wrong direction or more economically by reversing the connections between the transformer and the X-ray tube each time the secondary current changes its direction. This can be effected by a rotating commutator if the latter can be driven at exactly the right speed. The principle of the commutator will be understood from Fig. 40. The wires from the secondary coil are brought to two plates A and B placed at opposite ends of a diameter of a rotating insulated disc C. At the opposite ends of the diameter at right angles to this are two other plates D and E which are connected to the X-ray tube. The disc rotates between these four plates and carries four metal points, P, Q, R, S, joined together in pairs, as indicated in the figure, by conducting wires.

Suppose now that at the moment when the plate is in the condition shown in the figure the current in the secondary is flowing in the direction indicated by the arrow, so that A is the positive terminal and B the negative. Now A and D are connected by the conductor PQ, while E and B are connected by RS. Thus D becomes positive and E negative, and the current flows through the tube in the proper direction.

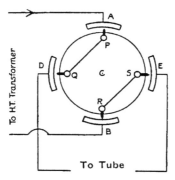

FIG. 40.—High tension commutator.

Now the secondary current is alternating with a frequency equal to that of the primary current, say, 60 per second. Thus, in 120th part of a second after the condition shown in the diagram, the current will be reversing its direction and A will now be negative and B positive. Suppose, however, that in this small time the disc has rotated through exactly one-quarter of a revolution, then P will be immediately opposite E and Q opposite A. Thus A and E are now joined, and thus E will still be negative and, similarly, D will still be positive. The current will, therefore, still flow from D to E through the tube, that is, in the same direction as before. Thus, by driving the disc at exactly the proper speed, the alternating current in the secondary can be converted into direct current in the tube circuit.

The disc can be rotated by what is known as a synchronous motor driven by the same alternating current supplied to the primary of the transformer. The rate of rotation of these motors is controlled by the periodicity of the current, and the motor is, therefore, always in step with the latter. Thus, when once adjusted, the rectification of the current is always complete.

60. COMPARISON OF TRANSFORMER AND INDUCTION COIL. —The relative advantages of the high tension transformer and the coil with break have given rise to much discussion, which is far from being settled. American radiologists seem to prefer the former, while the latter is more extensively used in this country.

On behalf of the transformer it may be stated that the output is larger than that of a coil, transformers capable of giving up to 10 kilowatts having been constructed. A kilowatt is 1000 watts, so that such a transformer would be capable of giving a current of 100 milliamperes at a potential difference of 100,000 volts. The large power available makes instantaneous radiography of almost any part of the body a possibility. The efficiency of the transformer, that is to say, the ratio of the power in the secondary to the power in the primary is also greater than that of a coil. This, however, is not a matter of much importance, as the cost of electrical power is not a large item in comparison with the other expenses of a radiographical department. The current supplied is unidirectional, there being no inverse current. The absence of an interruptor (an instrument which occasionally gives trouble to less experienced practitioners) is also an obvious advantage. On the other hand, should the moving parts of the transformer go wrong, the services of a skilled electrical engineer will be required to put them right.

On behalf of the induction coil it is claimed that, while the electrical efficiency of the coil itself may be less than that of a transformer, the current given by the coil is much more efficient in producing X-rays of the quality needed for radiographic and therapeutic purposes. On breaking the circuit of an induction coil the induced E.M.F. rises almost instantaneously to its maximum value. It remains at this value during practically the whole time the current is flowing, and then falls almost instantaneously to zero, remaining at that value until the next interruption. This is shown diagrammatically in Fig. 41, *a*. With the transformer, however, the current rises gradually to a maximum, and falls off equally gradually, rising again directly it has reached its zero value (Fig. 41, *b*). There is no period of rest, and a large part of the current is at a comparative low potential. Now the hardness of the X-rays produced increases with the voltage, so that, while in the case of the coil a large fraction of the rays are of the hard quality desired, in the case of the transformer there is a more considerable production of soft X-rays. The difference in the nature of the two discharges also results in a much greater heating of the tube for a given quantity of X-radiation emitted. Thus the tubes are more easily overheated, and certainly do not last so long with a transformer as with an induction coil. There cannot be any doubt that a current of the wave-form of Fig. 41, *a*, is far more efficient in producing X-radiation than that of the form of Fig. 41, *b*, and that it is the initial " kick," so to speak, given to the tube by the current which produces the radiation we particularly require. Improvements in coil design should aim at accentuating this initial impulse.

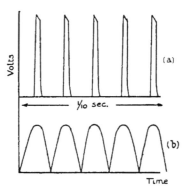

FIG. 41.—Current produced by (*a*) induction coil, (*b*) transformer

The flexibility of the induction coil, that is to say, the ease with which it can be adjusted to give the best results under different conditions as to the hardness of the tube, quantity of radiation required, etc., is greater than that of the transformer, although the latter has been greatly improved in recent years.

The balance of advantage would appear to be at present in favour of the induction coil, except when very heavy discharges are required for instantaneous work. Both instruments are undoubtedly capable of further improvement, and it may not be beyond the powers of electrical engineering to produce an instrument combining the best features both of the coil and the transformer.

CHAPTER VIII

61. COILS AND TUBES.—These are the vital parts of the installation, the rest of the apparatus being more or less subsidiary. Assuming that a coil is to be used and not a transformer, a good coil by a well-known maker should be selected. Economy in this vital part of the installation is fatal. The coil should be capable of giving a 16-inch spark in air, and of producing a current through an 8-inch spark gap of at least 30 milliamperes, when used with a suitable break. The secondary should be built in sections, and it will be convenient if the primary coil is arranged so that different primary voltages may be employed. A mercury interruptor either of the rotating jet type or the centrifugal type will be required, and for rapid work an electrolytic break should also be provided. If a high tension transformer is installed, a 4-kilowatt machine will be found sufficient for all practical purposes, but a beginner will probably find an induction coil more convenient.

A set of six X-ray tubes will be needed if the work is at all extensive. This will prevent the necessity for overworking any one tube, and by nursing the tubes up to different degrees of hardness, there should be always one tube suitable for any particular class of work. If, however, the Coolidge tube (§ 47) is used, this variety of tubes will not be necessary, as the hardness of the tube is completely under control from the switchboard and the Coolidge tube is not easily overrun. In this case one or two spare tubes, in case of accident, are all that will be required.

If a Coolidge tube is not employed, high tension rectifiers either of the mechanical type, or of the valve type, will be required if an induction coil is the source of current. It is usual to employ two valves in series in the secondary circuit, and a spare pair should be kept in reserve in case of accident.

62. SOURCES OF ELECTRICAL SUPPLY.—The most convenient source of supply, when available, is that provided by the Electric Supply Companies. This may be either direct or alternating in character, and may have almost any voltage from 100 to 300 volts, according to the local conditions and the preferences of the engineer in charge. The voltage, nature of the current (whether alternating or direct, and if the former, the periodicity also) must be ascertained before ordering any apparatus.

An induction coil requires direct current, so that if the supply is alternating a motor generator will be required to transform the alternating into direct current. An alternative method is to install a battery of accumulators and to charge them from the alternating circuit by means of a chemical rectifier, as described in § 22. A battery of accumulators sufficiently large to supply the currents now in vogue is expensive and requires much attention. The method is not to be recommended, except in the case of quite small installations which are not in very frequent use.

High tension transformers, on the other hand, require an alternating supply. Thus, if the supply is direct, a motor generator will be required to change the direct current into alternating current. It is possible that the nature of the current available may be the deciding factor between the use of a coil or of a transformer.

Where no electrical supply is available, by far the best way out of the difficulty is to install a small dynamo worked by a petrol engine. Several very satisfactory sets of this kind have been placed on the market for the lighting of country houses, and they are found to work very satisfactorily with a minimum of skilled attendance. In the field service outfits used during the late war small dynamos were arranged so that they could be geared on to the engine of a motor car, and the arrangement seems to have given satisfaction.

63. REGULATION OF THE CURRENT AND VOLTAGE.—The current supplied to the coil can be regulated by means of

rheostats (§ 20), used either in series or in parallel, or both. If both methods are employed we may regard the shunt rheostat as regulating the voltage across the coil, while the series rheostat controls the current. As, however, a decrease in voltage will produce a corresponding decrease in current, either type can be used to reduce the current through the apparatus.

The various devices for controlling the apparatus are generally collected in a switchboard. This, in addition to the actual control rheostats, should also be furnished with an ammeter and a voltmeter to measure the actual current and voltage supplied to the coil. It will also require a switch for making the actual exposures, that is, some device for making and breaking the coil circuit as required.

Further elaborations can be added to suit the tastes of the purchaser or the maker. A separate rheostat is often used to control the current passing through the motor of the mercury break, and so to alter its rate of revolution, while special keys may also be included to allow of a rapid change from a mercury break to an electrolytic break, or to increase the number of points in use in the electrolytic break if the latter is furnished with more than one.

If the apparatus is sufficiently powerful to permit of very short exposures (less than one second), an automatic exposing switch will be required which can be arranged so that the primary current is broken automatically after passing for a definite fraction of a second. Automatic switches for giving longer exposures can also be obtained, but are quite unnecessary.

If the operator has to work the installation without assistance, it may be convenient to have the switchboard mounted on a small table running on castors. The wires connecting the switchboard to the rest of the apparatus are, however, liable to get tangled, besides acting as a trap to the feet of the unwary. It is better, therefore, to have the switchboard permanently mounted on the wall in some convenient place, and this will generally be done when the operator can command the services of an assistant. It will be a great convenience if the lighting

of the room is controlled from the same board. The board
should be furnished with a shaded light, preferably a red light,
to enable the instruments to be read and the regulating handles
adjusted while the rest of the room is in darkness.

A diagram of the connections for a simple switchboard,
having shunt and series regulators and exposing switch, is
given in Fig. 42, and will sufficiently illustrate the principles

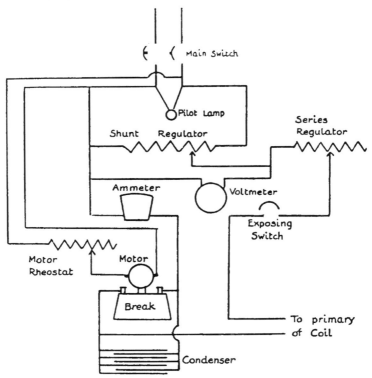

FIG. 42.—Diagram of connections for an induction coil.

involved. The actual details in any case will vary with the
particular make of board employed.   It should be remembered
that any elaborations, while adding to the convenience of
working, will also make it more difficult to localise any fault
in the circuit which may arise. The greatest simplicity
compatible with efficiency is to be recommended in all radiographic apparatus.

In the diagram the voltage required for the coil is selected
by means of the shunt rheostat, the current being also regulated

by a series rheostat. The actual current flowing is indicated by an ammeter in series with the coil, while the voltage is recorded on the voltmeter, which must be placed in parallel with the coil. The voltmeter is conveniently connected on the " main " side of the exposing switch. It then records the voltage of the circuit, even when the exposing switch is not closed. It will be noticed that the " main " current flows

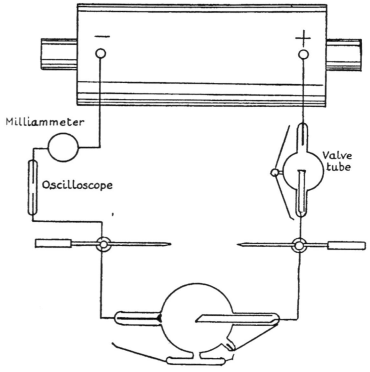

Fig. 43.—Diagram of connections for secondary circuit.

through the shunt resistance whether the current is being supplied to the coil or not. It can only be stopped by taking off the main switch. The pilot lamp serves as a visible indication that the main current is flowing. It can also be used to illuminate the switchboard.

The motor circuit is also shown in the diagram. A rheostat for adjusting the current through the motor is often added as shown. The condenser is connected in parallel with the mercury interruptor, as shown in the diagram.

The connections for the secondary circuit are sufficiently indicated in Fig. 43.

64. COUCH AND TUBE BOX.—Some sort of couch will be required for the screen examination of the patient, and for taking radiograms. The couch consists of a wooden table, the

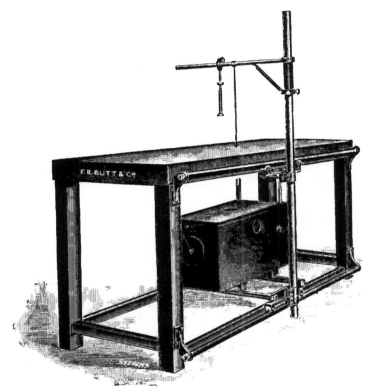

FIG. 44.—X-ray couch.

top of which is made of some substance comparatively transparent to the rays. Canvas stretched tightly will serve, but the sagging of the canvas under the weight of the patient is inconvenient. Three-ply wood or five-ply wood is much better, as it provides a comparatively flat surface for working on. With the quality of the radiation usually employed for radiography, the absorption in the wood is quite negligible.

Some arrangement will also be required for holding the X-ray tube, and for bringing it rapidly and conveniently under any desired part of the table. This can most easily be effected by clamping the tube inside a box running on castors on a small track placed at right angles to the length of the couch. The track itself is mounted on wheels, which run on rails placed parallel to the length of the table. The box has thus two movements, one at right angles and the other parallel to the length of the table, and can thus easily be brought under any desired part of the table. If the bearings are good the box can easily be moved with one hand. Each of the tracks should be furnished with scales to measure the distance which the box is moved in both directions. These will be required for stereoscopy and localisation. A convenient type of couch is shown in Fig. 44.

The box itself should be lined with at least 4 mm. of lead-impregnated rubber in order to protect the operator from the action of the rays. It is now well known that the continued action of X-rays sets up dermatitis of a very intractable kind, and efficient protection must always be provided. This is most easily and effectively done by cutting off the rays at the source and only allowing those actually employed for the work in hand to emerge. When the tube is employed under the couch as described, there is no difficulty in lining the box with sufficient absorbing material to stop practically the whole of the radiation, the extra weight, though very considerable, being of little importance if ball bearings are used for the wheels. Many of the tube boxes actually supplied are, however, far from being efficient. The box should always be tested with a fluorescent screen, and if there is any appreciable luminosity produced by the rays coming through the sides of the box, additional layers of lead rubber should be added.

When the tube has to be used above the couch, as may be sometimes necessary, it can be carried on one of the numerous varieties of stands which are catalogued by instrument makers. These generally have some sort of protection for the front of the tube, but, owing to the great weight of lead which would be required to cut off the rays completely, it is just as well to recognise that the protection in this case is never really efficient, and other methods of protection must be resorted to. Whenever possible the tube should be employed in its box below the couch.

65. USE OF DIAPHRAGMS.—The aperture in the box through which the rays emerge is fitted with an adjustable diaphragm to limit the size of the emerging beam, and to cut off as far as possible all unnecessary radiation. We have seen that in order to obtain a sharp shadow it is necessary that the rays should emanate from a single point. This is arranged for in the tube by focussing the beam of cathode rays on the target. The primary rays, however, when they fall upon the glass walls of the tube set up secondary radiations which proceed in all directions from the glass. These secondary rays, coming from an extended source, cause considerable fogging and loss of definition in the radiogram. An opaque diaphragm placed close to the tube will screen off this secondary radiation, except from the portion of glass just beneath the diaphragm, and thus materially improve the definition of the resulting picture. This is illustrated by Fig. 45. The scattered radiation from the walls of the tube is indicated by the broken lines in the figure. In the absence of the stop S these rays would reach the plate P and obviously produce a considerable degree of general fogginess.

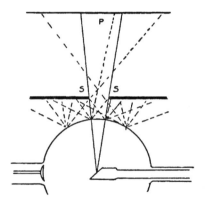

FIG. 45.—Effect of a diaphragm.

The stop S, however, stops a large proportion of these rays, and the smaller it is, and the nearer to the tube, the greater will its effect be. In practice the improvement produced is very striking, small foreign bodies, for example, such as needles, which may be quite invisible when the aperture is large, showing up plainly when the diaphragm is nearly closed.

The diaphragms supplied are often of the " iris diaphragm " type. This type, however, is not satisfactory, for two reasons. In the first place the aperture is always practically a circle, and it frequently happens that a circle is not the figure which fits most closely the part we wish to radiograph. In the second place the metal sectors of the iris diaphragm must necessarily be kept fairly thin, and they are very rarely sufficiently opaque

to the rays to form a really efficient screen. A much better
kind consists of two pairs of plates of lead, which are arranged
so that they can be moved nearer together or further apart by
means either of a screw or a lever motion. Each pair can be
adjusted independently, so that the aperture left between the
plates may be a rectangle of any suitable dimensions. The
plates can be made sufficiently thick to cut off practically the
whole of the radiation falling on them, and thus form an
efficient " stop." The only objection to this pattern is that it
requires two movements for its adjustment, while the iris, of
course, is adjusted by a single movement of the lever. In some
cases diaphragms in the form of a tube are used. These tubular
diaphragms cut off rather more of the secondary radiation than
the flat type. They are not adjustable, and it is doubtful
whether they produce any better result than the flat type if the
latter is mounted close enough to the tube.

In radiography the diaphragm should always be closed down
to the smallest area which will include all the part of the picture
which it is required to record. A radiogram of a special part of
the subject taken with a very small aperture will often reveal
details which are too blurred to be visible on a radiogram taken
with a large aperture. The adjustment of the aperture to the
size and place required can be effected by a preliminary
observation with a fluorescent screen.

66. SUBSIDIARY APPARATUS.—Of the large amount of subsidiary apparatus which fills up the makers' catalogues, very
little is really indispensable, though much may be useful in
special cases. For inhibiting the motion of the patient during
a time exposure various forms of clamps may be attached to the
X-ray table. A number of sandbags of suitable weight and size
will, however, answer the purpose equally well, besides being
more adaptable to individual necessities. Numerous ingenious
plate holders, for fixing the photographic plate in position during
exposure, are also available, but a few empty cardboard boxes,
with a sandbag or two, will often prove more satisfactory in
holding the plate in exactly the required position. For special
classes of work, of course, apparatus specially designed for that
class of work will be of great assistance. It is advisable to wait

until the necessity arises before choosing the apparatus. The radiographer will then be in a better position to discriminate between the various types on the market.

In the radiography of the abdominal region, the thickness of substances to be traversed by the rays can be materially diminished by the use of a suitable compressor. This not only reduces the exposure, but, what is still more important, reduces the intensity of the scattered radiation, and thus greatly improves the definition of the radiogram. Compressors of various shapes may be obtained either to clamp on to the X-ray couch, or to the overhead tube stand. The former method is preferable, unless the tube stand is exceptionally rigid, as any slight movements transmitted from the patient to the stand will shake the X-ray tube, and thus blur the resulting radiogram. When the exposure is made with the tube below the couch, a fair amount of compression can be obtained by placing a small air cushion between the patient and the table. The weight of the patient produces the compression, and can be assisted by a few sandbags if necessary. The results, however, are not so good as those obtained with a proper compressor.

CHAPTER IX

MAKING THE RADIOGRAM

67. EXPOSING THE PLATE.—If the tube is used below the couch the making of the radiogram is a comparatively simple matter. The rays are turned on, and, by means of a fluorescent screen examination, the exact part to be radiographed is located, and the diaphragm is closed down until only the particular region required is illuminated by the rays. A photographic plate, enclosed in a couple of light-tight wrappers, is then substituted for the screen, with the film side downwards, that is, facing the tube. X-ray plates can be purchased ready wrapped, and the film side is always indicated on the wrapper. Usually the plate is simply placed upon the patient and steadied by means of one or more sandbags placed upon it. Some couches are provided with elaborate plate holders for holding and clamping the plate. These are sometimes a convenience, but generally are more trouble than they are worth. The patient is then reminded to keep still and to hold his breath (if possible), and the exposure made by means of the exposing switch, being timed by means of a clock with a large centre-seconds hand.

It is very convenient if the number of plates dealt with is at all large to number the plates in succession. The number of the plate can be recorded on the plate itself by means of small lead numerals, which can be attached to the front of the plate by a strip of adhesive plaster. The numerals, being opaque to the rays, cast their shadows on the plate, and thus appear in white when the plate is developed. It is also desirable to record on the plate, in the same way, which is the patient's right and which the left. It is often more difficult than one would easily imagine to determine this from the radiogram itself. It takes very little time, and obviates all uncertainty, if a little lead R is attached to the side of the film nearest the patient's right.

68. TIME OF EXPOSURE.—The correct exposure of a plate is a matter requiring some little experience to obtain the best possible result. The time required depends on the following factors :—·

(1) *The Intensity of the Radiation given out by the Tube.*— For a given tube and a given spark gap, this will be approximately proportional to the current passed through the tube, that is to say, to the readings of the milliammeter. It is therefore imperative that a milliammeter shall be included in the secondary circuit. Different tubes undoubtedly differ somewhat in their efficiency, that is to say, in the intensity of the radiation given off with a given current. The variations, however, are not very great, if the tubes are of a reliable make. We can therefore, as a rough approximation, measure the exposure in milliampere seconds, that is to say, by the product of the secondary current in milliamperes into the duration of the exposure in seconds. Thus, if the correct exposure is stated as being 50 milliampere seconds, this implies that the plate will be correctly exposed if a current of 5 milliamperes is passed through the tube for 10 seconds, 10 milliamperes for 5 seconds, and so on.

(2) *The Distance between the Plate and the Anticathode.*— Since the intensity of the rays reaching the plate varies inversely as the square of the distance, it is obvious that if we double the distance between the plate and the source of the rays, we shall have to expose the plate for four times as long in order to produce the same action. The standard distance between the plate and anticathode is 50 cm. This gives a reasonably short exposure without producing undue distortion in the radiogram (see § 73). If for any reason the standard distance is varied, due allowance must be made in calculating the exposure. Thus at a distance of 35 cm. the exposure necessary is practically half the normal.

(3) *The Thickness of the Part to be Radiographed.*—This may affect the exposure in two ways. In the first place, when the tube is used under the table and the plate laid upon the patient, it is obvious that the distance between the plate and the tube increases with the thickness of the part, and the exposure is thus increased. Again, as flesh is not completely transparent to the

rays, the amount of absorption of the primary beam increases with the thickness which it has to penetrate (§ 39). Thus the rays are considerably weakened by their passage through the body, and the exposure must be proportionately prolonged.

This effect obviously depends very largely upon the quality of the rays ; the softer the rays the greater the absorption. On the other hand, very hard rays produce comparatively little effect on the photographic plate in proportion to their intensity, and thus require a longer exposure to produce the same action. It is obvious, therefore, that for every part there will be some particular quality of the radiation which will produce the desired effect in the minimum of time. In general, the thicker the part to be radiographed, the harder should be the tube. On the other hand, the harder rays show less distinction between different kinds of absorbing substances, and radiograms taken with a hard tube are, therefore, lacking in contrast and detail, and are generally less satisfactory for purposes of diagnosis. It would be useless, for example, to expect to find a calculus in the kidney with the rays from a tube of 7-inch or 8-inch spark gap. Thus, in spite of the great thickness of this region, examinations for calculus must always be made with a comparatively soft tube. While it is obviously impossible to give any hard and fast rule, it may be reckoned that each additional degree of softness of the tube on the Wehnelt scale will increase the exposure by about 30 per cent.

(4) *Make of Plate employed.*—Different brands of X-ray plates differ considerably in their speed, that is to say, in the exposure they require, some makes of plate requiring twice as much exposure as others. Various investigations on the speed of the plates most commonly employed have been published from time to time, but a given brand of plate may be gradually improved by the maker, or, on the other hand, may, for unknown reasons, show a gradual deterioration, so that no great stress can be placed on the results. The matter is one which the radiographer must test for himself.

The conditions for correct exposure are thus very complex. The following table will, however, serve as a rough guide to beginners in the art, and may be corrected by experience to suit the particular apparatus and mode of working. In working from below the table, the distance from the anticathode to the plate will vary with the thickness of the part to be radiographed. The table is drawn up for a couch in which the distance from the anticathode to the surface of the couch is 35 cm. If this distance is reduced to 25 cm., as may easily be done if required, the exposures would be approximately half those given in the table. The distortion would, however, be considerably increased, and, with the powerful discharges now available, any reduction in the distance is not to be recommended except in special circumstances.

The exposure required to produce a good radiogram even of a thick part is small compared with the permissible exposure to X-radiation. According to Hirsch, the limit of safety with a 5-inch alternative spark-gap tube at a distance of 14 inches from the patient is about 1,500 milliampere seconds. It is less for a softer tube and greater for a harder tube. Danger, however, might arise if a considerable number of radiograms were taken of the same part, or if the preliminary screening were unduly prolonged.

TABLE III.—EXPOSURE TABLE.

Distance from Anticathode to Nearest Surface of Part (i.e., to Table Top), 35 cm. (14 inches).

Object.	Penetration. Wehnelt units.	Milliampere seconds.
Hand, wrist	6	20
Forearm	6	45
Elbow joint	7	60
Shoulder joint	8	120
Thorax	8	120
Lumbar region	9	360
Calculus in kidney . . .	6–7	360
Calculus in bladder . . .	6–7	270
Pelvis	9	270
Hip joint	9	270
Femur	8	180
Knee joint	8	135
Lower leg	7	90
Ankle joint	7	75
Skull (lateral)	9–10	270
Skull (occipito-frontal) . .	9–10	360

69. USE OF THE INTENSIFYING SCREEN.—The exposures required may be much reduced by the use of what is known as an intensifying screen. This consists of a thin sheet of cardboard, one surface of which is coated with a thin layer of some fluorescent substance. The screen is placed with its coated side in intimate contact with the film side of the photographic plate, the rays passing through the screen on their way to the plate. The parts of the screen acted upon by the rays fluoresce more or less brightly, and the light emitted acts upon the adjacent photographic film, and thus adds its effect to the direct effect produced by the X-rays. The substance chosen should be one which emits fluorescence of considerable actinic quality, *i.e.*, white or bluish light. The light given out by a barium platinocyanide screen, though of considerable visual power, would produce hardly any photographic effect.

As the fluorescent light is given out in all directions it is evident that, unless the radiogram is to be badly blurred, the contact between the film and the screen must be of the closest. Special holders, known as cassettes, are used for holding the plate and screen in contact and protecting the former from daylight while the exposure is being made. The wrappings must, of course, be removed from the photographic plate before it is loaded into the cassette, the whole operation of loading and unloading being performed in the dark room. Great care must be taken to avoid scratching or marking the delicate surface of the screen, and to remove all dust both from the screen and from the photographic film, as any mark on the screen and every particle of dust will certainly be reproduced on the resultant negative.

The use of a screen materially shortens the exposure, and this is of great importance in the radiography of parts of the body where motion is continually going on, for example, in the case of the chest or abdomen. Owing to the inevitable slight spreading of the light from the screen, and also to the grain of the screen, which is usually apparent on the negative, the radiograms have not quite the same quality as those obtained without its use, especially in the case of parts where complete immobility is easy to obtain. For the radiography of parts such as the heart, lungs or stomach, the screen is almost indispensable.

Most makers claim that their intensifying screens permit the exposure to be reduced to one-tenth of the normal ; in practice it is generally safer to assume that the exposure required with a screen will be about one-fifth of the exposure required without the screen.

70. DEVELOPMENT OF THE PLATE.—After exposure the plate requires to be developed to render the image visible, and fixed in order to make it permanent. The processes are identical with those employed in ordinary photography, and need not be described in detail. Almost any standard developer can be used. In the absence of any personal reference, that recommended by the maker on the box containing the plates should be selected. Metol hydroquinone is a good all-round developer. It has the advantage of being rapid in its action (a very considerable asset where a large number of plates have to be dealt with), and keeps fairly well in solution if the bottle is completely filled with the liquid and tightly corked. The temperature of the developer should be between 60° and 70° F.

The plate is placed film side upwards in the dish, and the developer flooded over it from a large measuring glass, care being taken that sufficient is used to cover the plate completely. The dish is then gently rocked until development is complete, which will be when the larger outlines of the subject can be seen appearing on the back of the plate. With normal exposure and a metol hydroquinone developer, the image should begin to appear in about fifteen seconds, and development will then be complete in from four to five minutes. If the image is slow to appear the plate has been under-exposed, and may then require to be left in the developer for as long as fifteen minutes. It is not worth while to leave it longer than this.

It is advisable to keep the developing dish covered during development, as even the ruby light of the dark-room lantern has a slight effect on the X-ray plates.

The same developer may be used for two or three plates in succession, but not more, as it becomes oxidised, partly by its action on the plates, and partly by exposure to the air.   It is not advisable to keep partly used developer for use another day. The cost of developer is small compared with that of the plates, and it is unwise to risk failure from the use of old and weak developer.

After development the plate should be rinsed under the tap and transferred to the fixing bath.   This may, with advantage, be one of the combined fixing and hardening baths, which harden the gelatine film, as well as remove all the unreduced silver salts from the film.   The negative should remain in the bath for at least five minutes after all traces of milkiness have disappeared from the film.   Fixing takes longer in the case of X-ray plates than in the case of ordinary plates, on account of the greater thickness of the gelatine film.

After fixing, the plate should be washed in running water for at least an hour, a washing tank being employed if the plates are numerous.   It may then be left to dry in a fairly warm, well-ventilated room.

If the plate is required for immediate use the drying may be hastened by first removing the surface water with a soft pad of cotton wool, and then soaking the plate for five minutes in methylated spirits.   If it is then removed and placed in a current of air, it will dry very rapidly.   This process is unnecessary unless the plate has to be sent away.   In the department itself the plate can be viewed in a viewing lantern while wet, and a diagnosis made.   In fact, the plate often seems to be at its best before it is dried.

71. PRINTING.—Prints can be made, if required, from the negative by any of the usual photographic processes.   No printing process, however, will reproduce all the detail which can be found on the negative, as no paper can record all the range of tones which can be obtained in a transparency. Special paper of the gaslight and bromide types is made by some firms for the purpose of making X-ray prints from the negatives and, on the whole, gives better prints than the ordinary gaslight or bromide papers.   In no case, however, is the copy of quite the same value as the original negative.

72. VIEWING THE RADIOGRAM.—The radiogram may be

examined by simply holding it up to a dull light. It is much more convenient, however, to employ a viewing lantern (Fig. 46) which consists simply of an opal glass illuminated from behind by an electric lamp, and fitted with a series of holders, having different sized apertures to fit the sizes of plates generally employed. A holder should be chosen of such size that no stray light can come out round the edges of the plate, and the plate is placed upon it with the film side outwards, that is, towards the spectator. The picture seen is then that which would have been seen by an eye placed at the anticathode of the X-ray tube. The negative will show more clearly if the rest of the room is in darkness, so that no stray light falls on the negative from the front.

The original plate is what photographers call a negative, that is to say, the parts of the plate acted upon by the rays will appear dark, those unacted upon will appear as clear glass.

Thus, in the radiogram of a limb, the portions of the plate, if any, which overlapped the limb altogether,

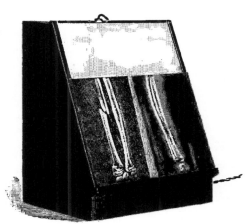

FIG. 46.—Viewing lantern.

and therefore received the full effect of the rays, will appear almost or quite dark. The flesh, which cuts off a small fraction of the rays, will appear somewhat less dense, and its outline should be easily traceable unless the plate has been much overexposed. The bones, which stop far more of the radiation, will appear much lighter, but should have sufficient density to show the structure. A metallic body embedded in the limb will, if at all thick, appear as clear glass. In general, any part where there is increased density will appear lighter than the surrounding parts, any part where the density is diminished, say, for example, a rarefied area in the bone, will appear darker than the surroundings. This is, of course, the reverse of the appearance seen on the fluorescent screen.

The appearance of a print will be exactly the reverse of the negative, the bones appearing dark, the flesh much lighter, and the background pure white. The relations between the negative and the print or " positive " are shown in Fig. 47, *a* being a reproduction of the actual plate, *b* the corresponding print from the same negative.

The radiographer will always prefer to make his diagnosis from the actual plate, that is, from the negative. It may be stated with confidence that no method of printing yet devised will enable us to reproduce on paper the full range of tones of an X-ray negative. Some of the detail is inevitably lost in the process. It is always possible, with practice, to make a print which will bring out any definite point on t h e skiagram, a n d where a print is required the printing will be adjusted to make clear the particular features on which the diagnosis rests. The diagnosis itself should,

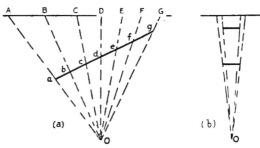

FIG. 48.—Distortion in the radiogram.

however, always be made in the first place from the negative.

The art of making the best print from a negative is one requiring experience. Books on photography may be consulted with advantage. The kind of " faking " which involves the use of pencil or brush on the negative is, however, not permissible in scientific work.

73. DISTORTION IN THE RADIOGRAM.—The radiogram or picture on the fluorescent screen provides only a kind of shadow-graph of the parts between the focus of the tube and the photographic plate or screen. It cannot be too persistently kept in mind that the " best of this kind are but shadows." In general, the radiogram will suffer not only from magnification, but also from distortion. Consider, for example, the case of the rod in Fig. 48, *a*. If O is the anticathode and AG the plate or fluorescent screen, the shadows of the points *a, b, c, d, e, f* will be formed along the prolongation of the lines joining these points to the focus O

FIG. 47a.—Reproduction of a radiogram of knee joint (normal) showing the appearance of the actual "negative," as viewed from *film* side.

FIG. 47b.—Reproduction of the print (positive) from the
negative of Fig. 47a.

of the tube, since the X-rays all travel in straight lines from the anticathode. Thus on the resulting radiogram the shadows of the points will appear at A, B, C, D, E, F, respectively. It will be seen at once that while the points a, b, c . . . are equally spaced this is by no means the case with their shadows. The spacings are all greater on the radiogram than they are in reality, that is to say, they are all magnified. In addition to this, however, there is also distortion, the distance between A and B, for example, being greater than that between C and D. The magnification is least in the neighbourhood of the central or normal ray, and greatest towards the edges of the plate. It is for this reason, among others, that the tube is usually centred so that the normal ray passes through the particular point in the object with which we are particularly concerned.

The magnification depends on the distance between the object and the plate, and between the focus and the plate. If the object is in actual contact with the plate, there will be no magnification. In accordance with the usual laws of shadows, the greater the distance of the object from the plate the larger will be the shadow which it casts (Fig. 48, *b*).

Increasing the distance between the plate and the anticathode decreases the magnification. Thus, if it were possible to have the distance between the plate and the anticathode 5 metres, the magnification of an object 10 cm. from the plate would only amount to about 2 per cent., which would be negligible. As, however, the intensity of the radiation varies inversely as the square of the distance, the exposure required would be one hundred times that at the standard distance (50 cm.), and either very powerful apparatus, or very prolonged exposures, would be required. This rather limits the application of this method of reducing distortion, to which the name of teleoradiography has been given.

Unless the focus of the tube is particularly sharp, the shadows of objects at some distance from the plate will also be blurred. It must be remembered that X-rays will be given off from every portion of the target which is struck by the cathode rays. Each point so struck will thus produce its own shadow of the object, and these shadows will overlap, producing a blurring of the fine detail.   The sharper the focus the sharper the shadow, so that if detail is required a fine focus tube must be employed. The focussing of the whole of the energy in the cathode beam on to a single spot, however, entails a greater strain on the tube, so that in cases where detail is not required (in the case of the examination of the stomach, for example), it is more economical to employ a tube in which the focus is diffuse.

The radiogram, being only a shadowgraph, gives no indication of the relative depths of the different shadows, except such as may be gathered indirectly from the sharpness of the shadow or from its size.   Two objects situated on the same ray will produce shadows in the same place on the plate.   For example, the shadow of the bullet in Fig. 51 would be seen overlapping that of the bone, and the plate would afford no indication as to whether the bullet was in front of, behind, or actually in the bone itself.   In the case considered, the question could be solved by taking a second radiogram with the tube at right angles to the first position.   This would at once show that the bullet was situated below the bone.   A pair of plates taken from two viewpoints at right angles to each other will often give all the information required.

An alternative method is to invoke the aid of stereoscopy. One or other of these methods should always be employed in cases when the actual position of the shadows is of any importance.   It is usually unwise to make a diagnosis from a single radiogram.

74. STEREOSCOPIC RADIOGRAPHY.—When we look at an object, using both eyes, the image formed on the fovea of the one eye is not exactly the same as the image formed on the fovea of the other, since the object is regarded from a slightly different standpoint by the two eyes.   The effect can easily be demonstrated by looking through a window at some object outside.   On closing first one eye, and then the other, the window frame appears to move across the background, showing that the image is different in each eye.   This difference between the images gives the impression of solidity.   When we look at a flat surface, a picture, for example, the two images are precisely the same, and thus, even though the perspective

and shading may be correct, we do not get the impression of relief.

This impression of solidity or relief can be obtained by means of the stereoscope. Two photographs of the same object are taken from two slightly different standpoints, say, some 6 cm. apart. The two photographs will then correspond with the images seen by the two eyes in binocular vision. The photographs are then placed side by side, so that each eye sees its own corresponding picture, and the images formed by the eye are made to fall on the corresponding points of the fovea so that they appear to overlap and fuse into a single picture. This can be done by " squinting " at them, but the process is not very comfortable. The superposition of the two images can be achieved more comfortably by optical means.

The ordinary stereoscope consists, in principle, of two prisms of the same angle placed edge to edge, as shown in Fig. 49. A prism produces a virtual image which is displaced towards the edge of the prism, and thus, on looking through the prisms, the two pictures are both seen to be shifted towards the centre. By

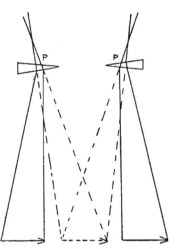

FIG. 49.—Principle of the stereoscope.

properly adjusting the distance between the prisms and the pictures these displaced images can be made to overlap exactly. They will then appear to fuse into a single picture, which, owing to the fact that the two separate pictures are slightly different, will give a remarkable sense of solidity or relief. In practice, the prisms are generally replaced by the two halves of a single convex lens, one half being turned round so that they come edge to edge. The displacement produced is similar to that produced by the prisms, and, in addition, there is a certain amount of magnification.

An exactly similar sense of relief can be obtained by combining two stereograms of the same object taken with the anticathode in two different positions at some short distance apart. The relative positions of the shadows in the two radiograms will be slightly different, and this difference is sufficient, when they are combined in a stereoscope, to give the impression of relief, exactly as in the case of a pair of stereoscopic photographs. It is thus possible to see at a glance the relative positions of the different objects appearing on the plates.

The prism stereosocope is not suitable for pictures much larger than 3 × 3 inches. As most of our radiograms are considerably larger than this, we must either reduce them to this size, by means of an enlarging lantern, a process involving some

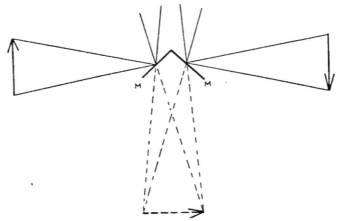

FIG. 50.—The Wheatstone stereoscope.

time and trouble, or use some other method of combining the two pictures. When the radiograms are required for use elsewhere, the former method is often employed. In the X-ray department itself the latter method is generally employed, as it enables the original radiograms to be used without any further preparation.

The Wheatstone reflecting stereoscope uses the properties of plane mirrors for producing the necessary superposition of the two pictures. Two plane mirrors are set at right angles to each other (Fig. 50), and the two pictures are set up, one on each side of the mirrors, so that the line joining the pictures makes an angle of 45 degrees with the plane of each of the mirrors. Each mirror forms an image of the corresponding picture which is exactly as far behind the mirror as the picture is in front of it, and displaced through exactly 90 degrees. The two images are therefore exactly superposed, and if the head is placed close to the angle between the mirrors, so that one eye looks into each mirror, the combined images will be seen standing out in stereoscopic relief.

If the actual negatives are to be used, as is always the case, they will require to be illuminated from behind, and for this purpose the instrument is generally made so that the two negatives are carried by two small viewing boxes, one on each side of the mirrors, and one of the carriers should be arranged to slide both in the vertical and the horizontal direction in its own plane, so that the two images can be brought into exact register. This adjustment will be required in case the two plates have not been placed in exactly the same position during the exposures. The best results are obtained when the distance between the mirrors and the radiograms is the same as that between the anticathode and the plate during the exposure, and in some of the more elaborate forms of instrument the carriers are arranged so as to slide towards or away from the mirrors. This adjustment is not, however, absolutely necessary.

To take a pair of stereoscopic plates, the tube is first centred in the usual way on the part required and is then displaced some measured distance, say, 3 cm. to the right, and the exposure is made in the usual way. The plate is then removed, and a second plate placed as nearly as possible in the same position as the first. The tube is then moved to exactly the same distance to the left of the normal position, that is to say, 3 cm. to the left of the original position, or 6 cm. to the left of the position during the first exposure, and a second exposure is made. It is important that the two exposures should be exactly the same, as otherwise one of the radiograms will be denser than the others, and combination will be difficult. The plates are then developed and fixed in the usual way. We have now two pictures of the same object taken from two slightly different standpoints.

The two plates are now placed in the stereoscope, film side

outwards (*i.e.*, facing the mirrors). Owing to the well-known lateral inversion of the image produced by a plane mirror, if we wish to see the resultant picture from the standpoint of the X-ray tube, we must place the left-hand exposure on the right of the mirrors, and the right-hand exposure on the left. The image will then give us the object in relief as seen from the point of view of the tube. If the left-hand exposure is placed on the left and the right-hand plate on the right, the image will still be seen in relief, but the relief will be reversed, that is to say, we shall get the picture which would be seen by an observer looking through the object from the opposite side to the tube.

It must be remembered that the skin does not cast a perceptible shadow, and as we shall generally want to get an idea of the depth of the various parts from the surface of the body, it is advisable to put a piece of lead wire on the skin, both front and back, to indicate its position. A little lead cross placed on the skin in contact with the plate is very convenient, as it will enable us to get the two images in exact superposition with the minimum of trouble.

The best displacement of the anticathode varies with the distance between the anticathode and the plate, and with the thickness of the part to be radiographed. The following table, calculated by Marie and Ribaut, gives the proper shift under different conditions, but quite good results can be obtained even when the movement is not exactly that given in the table.

Table IV.—Marie and Ribaut's Table.

Thickness of part to be radiographed in cm.	Distance of Anticathode to surface of the body.			Distance through which tube must be displaced between the two exposures.
	30 cm.	40 cm.	50 cm.	
4	5·4	8·8	13·5	
6	3·6	6·1	9·3	
8	2·8	4·1	7·3	
10	2·4	4·0	6·0	
15	1·8	2·9	4·3	
20	1·5	2·4	3·5	
25	1·3	2·1	3·0	

CHAPTER X

LOCALISATION OF FOREIGN BODIES

75. LOCALISATION OF FOREIGN BODIES.—A good pair of stereoscopic plates will give a good idea of the position of any foreign body in relation to the neighbouring parts of the body which are sufficiently dense to cast shadows on the plate. In cases of difficulty, where more exact information is required, resource must be had to some method of localisation. The two methods can be used with advantage to supplement each other, and the radiographer will generally make an exact localisation for his own satisfaction even when the surgeon prefers to work from stereoscopic plates.

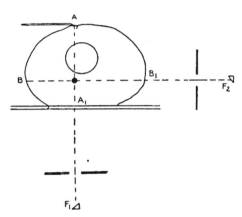

FIG. 51.—Localising by right angle views.

Much ingenuity has been spent on the problem of localisation, and some very elaborate and costly apparatus devised. This is necessary in special cases where the utmost accuracy is required, as, for example, in the case of foreign bodies in the eye. The principle employed is in all cases quite simple, and generally, where the position of the foreign body is not required to an accuracy of more than 2 or 3 mm., the process can be carried out with the simplest auxiliary apparatus.

76. LOCALISING BY RIGHT ANGLE MARKS.—In the case of a foreign body in a limb where two views at right angles to each other can be obtained, the exact position of the foreign body can be indicated very simply and rapidly. The limb is placed

in some standard position, which is carefully recorded, and the
tube is centred so that the shadow of the foreign body appears
on the fluorescent screen exactly in the centre of the aperture,
which will be closed as far as possible to facilitate this adjustment (Fig. 51).   A blunt probe is then inserted between the
skin and the screen, and moved until the shadow of the tip
of the probe is exactly on the shadow of the foreign body.
The screen is then removed and the position of the tip of the
pointer marked on the skin in indelible ink.   The foreign body
then lies on the straight line joining this mark to the focus of

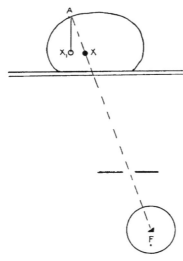

FIG. 52.—Error due to obliquity
of central ray.

the tube, and passing on its way
through the centre of the diaphragm.
If this central ray, as we may call it,
is accurately vertical, the foreign
body lies vertically below the mark
on the skin with the limb in the
given position.   The line of the ray
can be verified by inserting the probe
below the limb and making a second
mark at A_1.   The foreign body then
lies, in any case, on the line joining
AA_1.   Special probes or finders can
be obtained which mark their position automatically on the skin by
merely pressing a small bulb.   These
are very convenient, especially for
marking positions such as A_1.

The tube is then moved to a position F_2 exactly at right
angles to the first, and again centred on the foreign body.   The
latter then lies on the ray FB_1B.   The points B and B_1 can be
marked on the skin as before.   The foreign body must then
lie at the intersection of the lines AA_1 and BB_1.   If all four
points are marked as shown in the diagram, it is not necessary
that the two positions of the tube should be exactly at right
angles to each other, though it will generally be convenient to
have it so.   If only the points A and B are marked (as is often
the case) it is essential that the two central rays shall be exactly
at right angles, and that the central ray in the first case shall

be exactly vertical. Otherwise a very serious error may arise, as the operating surgeon would expect to find the foreign body vertically below the mark A (say at X_1), whereas if the central ray has been oblique, it may be considerably to one side of this line (Fig. 52).

Instead of turning the tube, which is not very convenient in practice, the limb itself may be turned through a right angle. The placing of the limb requires to be done with care and accuracy in order to obtain a true antero-posterior, and a true lateral view. Care must also be taken that the limb is turned as a whole, and not merely twisted, as in the latter case there might be a considerable displacement of the foreign body itself relative to the skin markings. This is a possibility which must always be borne in mind in all localisations.

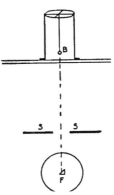

77. Finding the Vertical Ray.—It is generally important in most methods of localising that the central ray shall be accurately vertical. This adjustment will have to be made afresh every time the tube is changed. Apparatus for this purpose can be bought, or can be constructed in a few minutes. A metal or cardboard cylinder (such as a tobacco tin with the ends knocked off) has two wires stretched tightly across one

Fig 53 —Method of finding the vertical ray.

end to form a cross. From the centre of the cross is hung a small plumb bob by a cotton thread. The bob will thus always be vertically below the centre of the cross. The apparatus is placed on the couch in a vertical position, a small fluorescent screen placed over the wires, and the tube box brought below the apparatus, with the aperture nearly closed. The position of the tube in the box is then adjusted until the shadow of the bob appears exactly on the centre of the cross wires, and at the same time exactly in the centre of the diaphragm. Since the plumb bob is vertically below the cross, the central ray through the diaphragm must then be exactly vertical (Fig. 53).

78. Localisation by Triangulation. — The method of § 76 is only of service in the case where the foreign body is

in a limb.   In other positions, or even in the case of a limb if
very accurate results are required, resource must be had to
other methods.   Of these the most important is that of
triangulation.

Suppose that X (Fig. 54) is the foreign body, whose position
is to be localised.   The tube is centred on the foreign body,
care being taken that the central ray is vertical, and the
position of the shadow is marked on the skin in the usual way.
The foreign body is then vertically below the point A. Suppose

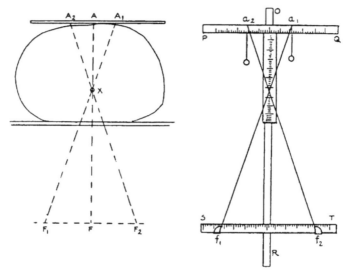

FIG. 54.—Localising by triangulation.

now that the tube is moved some distance to the left of this
position, say to F_1.   The shadow will move to A_1, where F_1XA
is a straight line.   The tube is now moved to a position as far
to the right of F as it was previously to the left, that is to a
position F_2 where FF_2 is equal to FF_1.   The shadow will move
to a corresponding position A_2.   The foreign body X is then
at the intersection of the lines F_1A_1 and F_2A_2.   If AF, A_1A_2,
and F_1F_2 are measured, this position can be calculated.

Since the triangles XA_1A_2 and XF_1F_2 are similar—

$$\frac{XA}{XF} = \frac{A_1A_2}{F_1F_2}$$

$$XF = \frac{F_1F_2}{A_1A_2} . XA$$

But $$XF + XA = AF$$

$$\therefore \quad XA + \frac{F_2}{A_1A_2} \cdot XA = AF$$

$$XA = \frac{AF \cdot A_1A_2}{A_1A_2 + F_1F_2}$$

In practice the calculation can be avoided by reconstructing the problem on a localiser, using threads to represent the rays. Suppose PO (Fig. 54) is a horizontal scale fixed at the top of a long vertical scale OR. A second horizontal scale ST is arranged to slide on the vertical scale. The distance between the two horizontal scales is made equal to the distance AF between the focus of the tube and the plate or screen on which the measurements were made. Two threads, a_1f_1 and a_2f_2, are fastened at their lower ends to carriers sliding on the horizontal scale ST, while their other ends hang over the top scale PO, the strings being kept taut by means of small weights attached to their ends. The carriers f_1, f_2 are set on the lower bar so that their distances from the centre are equal to the distances FF_1 and FF_2 of the tube during the exposure. Similarly, the upper ends are moved along the upper bar until their distance from the centre corresponds with that of the corresponding shadows. The position of the threads a_1f_1 and a_2f_2 then corresponds exactly with that of the rays A_1F_1 and A_2F_2 in the experiment. The foreign body is thus obviously at the point of intersection of the threads, and the distance of this point below the centre of the crossbar, which represents the mark A on the skin, can be read off directly on the vertical scale.

By far the most accurate way of determining the distance through which the shadow moves is by making two exposures upon the same plate with the focus in two different positions. After centreing the tube below the foreign body, and marking the ition on the skin, a plate is placed over the part and firmly xed in position, either by clamps or sandbags. Exposures are made with the tube, say 3 cm. to the left, and then 3 cm. to the right, of the central position, the plate remaining fixed in position during the whole process. Two separate shadows thus appear on the plate when developed, and the distance between them can be measured off with a pair of dividers and a graduated scale.

If preferred, two separate plates may be exposed, one for each position of the focus. In this case a pair of cross wires must be fastened on the skin in contact with the plate. As the shadow of the cross wire, which is in contact with the plate, does not change its position, it serves as a landmark enabling the plates to be brought into register after development. For example, if the shadow is found to be 2 cm. to the right of the wire in one plate and 2 cm. to the left on the other, the total shift of the shadow would be 4 cm. The method is obviously much more liable to error than that in which the two shadows appear on the same plate. It has the advantage that if the shift of the tube has been suitably chosen, the pair of plates form a stereoscopic pair, and can be viewed in the stereoscope.

79. MACKENZIE-DAVIDSON LOCALISER. — The Mackenzie-Davidson Cross Thread localiser acts on precisely the principle already described, the only difference being that by building up the thread model in three dimensions, instead of in two only, the preliminary screening can be avoided.

The plate-holder is furnished with a pair of cross wires, and by mechanical means the tube is centred vertically above the cross. The two exposures are made exactly as in the previous case, and after development the radiogram is placed so that the cross coincides with cross lines ruled on the table of the localiser. The centre of the cross is then vertically below the centre of a short cross bar which is carried by a vertical scale. The height of the cross bar above the plate is adjusted to be equal to the distance of the anticathode from the photographic plate during exposure, and then represents the line along which the focus of the tube was moved. Two threads are then stretched from two nicks on the cross bar at a distance from each other equal to the distance the tube was moved, and lead to the central point of the corresponding shadows on the plate. The two threads will then represent the two central rays by which the shadows were formed, and their point of intersection will be the position of the foreign body. The vertical distance of this point from the plate and its horizontal distance from each of the cross wires can then be measured. The position of the cross wires will have been marked on the skin immediately after the exposure, and the point of the skin which is vertically above the foreign body can thus be found and marked on the skin. The method has the advantage that it can be applied to cases where the foreign body cannot be seen on the fluorescent screen. For foreign bodies sufficiently large to be readily visible it seems to present no advantages over the simpler method.

80. LOCALISING WITH THE FLUORESCENT SCREEN.—The method of triangulation can also be applied with the fluorescent screen. For this purpose the screen may be fitted with two pointers which slide on a graduated scale at the side of the screen. Some means will also be required for keeping the screen fixed in one position during the observations. One of the pointers is made to coincide with one of the shadows of the foreign body. The tube is then moved a measured distance, and the second pointer is moved until it coincides with the new position of the shadow. The distance through which the shadow has moved can then be read on the scale.

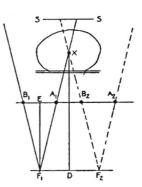

FIG. 55. — Screen method of localisation.

A much better method, if screen localisation has to be employed, is the following. A special diaphragm is mounted above the tube box, and two wires are stretched across it at right angles to the direction in which the tube moves on its rails. It is generally convenient to move the tube at right angles to the length of the table, in which case the wires will be parallel to the length of the table. The distance between the wires should, for convenience, be some definite fraction of the distance of the wires from the anticathode, say one-third or one-quarter, and this adjustment requires to be made accurately, as the accuracy of the result depends upon it.

The tube is then placed so that the shadow of one of the wires intersects that of the foreign body, and the position of the tube box is read on the scale attached to the slide on which it moves. The tube is then moved at right angles to the direction of the wires until the shadow of the second wire intersects that of the foreign body. There is no need to keep the screen fixed or to know its distance from the tube, as the position of the screen obviously makes no difference to the adjustments. The distance which the tube has been moved is then observed.

Let F_1 (Fig. 55) be the first position of the focus of the tube, A_1, B_1 the corresponding positions of the wires, and X the foreign body. Then when the shadow of X coincides with that of A, XA_1F_1 must be a straight line, as drawn. Similarly, the second position, in which the shadow of X coincides with that of B, must be XB_2F_2. Now, as the focus F and the wires A, B move together, B_1F_1 is parallel to B_2F_2, and A_1F_1 is parallel to A_2F_2. Thus, the triangles XF_1F_2 and $F_1B_1A_1$ are similar. Thus—

$$\frac{\text{Height of triangle } XF_1F_2}{\text{Base of triangle } XF_1F_2} = \frac{\text{Height of triangle } F_1B_1A_1}{\text{Base of triangle } F_1B_1A_1}$$

$$\therefore \quad \frac{XD}{F_1F_2} = \frac{F_1E}{A_1B_1}$$

$$XD = \frac{F_1E}{A_1B_1} \times F_1F_2$$

$$= \frac{\text{distance of wires from focus}}{\text{distance between wires}} \times \text{shift of the tube.}$$

The ratio of the distance of wires from the focus to the distance apart of the wires can be adjusted to be some definite whole number, say 3, for example. In this case we have—

Distance of foreign body from focus of tube

$$\coloneqq 3 \text{ times the shift of the tube.}$$

Thus, when once the position of the diaphragm has been properly adjusted, the localisation can be carried out with the greatest speed and convenience. The special diaphragm can easily be mounted on a swivel, so that it can be turned out of the way of the rays when not required for use.

It must be noted that the calculation gives the distance of the foreign body from the anticathode of the tube. The distance of the surface of the couch from the focus should be known. If this is subtracted from the result, the difference will give the distance of the foreign body above the surface of the couch. Its depth above or below any mark on the skin can then easily be ascertained. This is probably the best of the screen methods of localising. In no case, however, will a screen method give the same degree of accuracy as can be obtained by photographic methods.

INDEX

(The references are to pages.)

PRINTED IN GREAT BRITAIN BY THE WHITEFRIARS PRESS. LTD., LONDON AND TONBRIDGE.

Eastman
Dupli-Tized X-Ray Film

The Film consists of a special emulsion coated on both sides of a thin, flexible transparent base. This method of double coating, greatly reduces the time for exposure. Two images are secured—super-imposed—which give good density and delicate gradation. In addition, the films are light, unbreakable, easy to file and take up less room than glass plates.

When you have failed with glass plates you can succeed with Eastman Dupli-Tized X-Ray Film.

The Kodak
System of Constants

Use Dupli-Tized Film and the Kodak System of Constants from the start. The System is a scientific principle worked out to its simplest form. Call at Kingsway—we can teach you the photographic side of Radiography in half-an-hour.

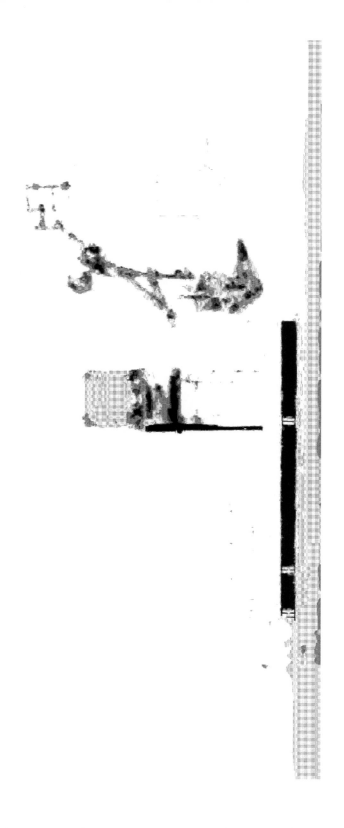

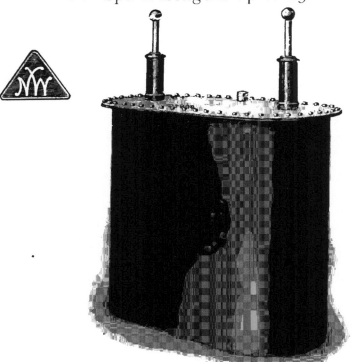

THE RADIOSTAT
High-Tension Transformer Oil Immersed

FOR COOLIDGE OR GAS TUBES

For fuller particulars apply for leaflet (in preparation) to the Makers.

Our Standard Model Radiostat is constructed in sizes up to 15 K.V.A. for either D.C. or A.C.

The Model illustrated shows the Standard type for A.C. mains with control for Coolidge or Gas Tubes. The Unit is also supplied with a separate Switch-table Control.

Of our usual sound design and robust construction, with exceptionally high output.

A. E. DEAN & CO.,

MANUFACTURERS of X-RAY and ELECTRO-MEDICAL APPARATUS
of the HIGHEST QUALITY

Leigh Place, Brooke Street, Holborn, London, E.C. 1